Practical Discrete Mathematics

Discover math principles that fuel algorithms for computer science and machine learning with Python

Ryan T. White

Archana Tikayat Ray

BIRMINGHAM—MUMBAI

Practical Discrete Mathematics

Copyright © 2021 Packt Publishing

All rights reserved. No part of this book may be reproduced, stored in a retrieval system, or transmitted in any form or by any means, without the prior written permission of the publisher, except in the case of brief quotations embedded in critical articles or reviews.

Every effort has been made in the preparation of this book to ensure the accuracy of the information presented. However, the information contained in this book is sold without warranty, either express or implied. Neither the authors, nor Packt Publishing or its dealers and distributors, will be held liable for any damages caused or alleged to have been caused directly or indirectly by this book.

Packt Publishing has endeavored to provide trademark information about all of the companies and products mentioned in this book by the appropriate use of capitals. However, Packt Publishing cannot guarantee the accuracy of this information.

Group Product Manager: Ashwin Nair

Publishing Product Manager: Pavan Ramchandani

Senior Editor: Hayden Edwards

Content Development Editor: Aamir Ahmed

Technical Editor: Deepesh Patel

Copy Editor: Safis Editing

Project Coordinator: Kinjal Bari

Proofreader: Safis Editing

Indexer: Manju Arasan

Production Designer: Vijay Kamble

First published: January 2021

Production reference: 1210121

Published by Packt Publishing Ltd.
Livery Place
35 Livery Street
Birmingham
B3 2PB, UK.

ISBN 978-1-83898-314-7

www.packt.com

To my parents, for their endless support, and the teachers and mentors, formal and informal, who inspired me.

– Ryan T. White

To my parents and my sister, Abhipsha, for their unending love and support. To Amma and Appa, for being the coolest in-laws. Finally, to my best friend, husband, and the love of my life, Anirudh, for always having my back and believing in me.

– Archana Tikayat Ray

Packt.com

Subscribe to our online digital library for full access to over 7,000 books and videos, as well as industry leading tools to help you plan your personal development and advance your career. For more information, please visit our website.

Why subscribe?

- Spend less time learning and more time coding with practical eBooks and Videos from over 4,000 industry professionals
- Improve your learning with Skill Plans built especially for you
- Get a free eBook or video every month
- Fully searchable for easy access to vital information
- Copy and paste, print, and bookmark content

Did you know that Packt offers eBook versions of every book published, with PDF and ePub files available? You can upgrade to the eBook version at packt.com and as a print book customer, you are entitled to a discount on the eBook copy. Get in touch with us at customercare@packtpub.com for more details.

At www.packt.com, you can also read a collection of free technical articles, sign up for a range of free newsletters, and receive exclusive discounts and offers on Packt books and eBooks.

Contributors

About the authors

Ryan T. White, Ph.D. is a mathematician, researcher, and consultant with expertise in machine learning and probability theory along with private-sector experience in algorithm development and data science. Dr. White is an assistant professor of mathematics at Florida Institute of Technology, where he leads an active academic research program centered on stochastic analysis and related algorithms, heads private-sector projects in machine learning, participates in numerous scientific and engineering research projects, and teaches courses in machine learning, neural networks, probability, and statistics at the undergraduate and graduate levels.

Archana Tikayat Ray is a Ph.D. student at Georgia Institute of Technology, Atlanta, where her research work is focused on machine learning and **Natural Language Processing** (**NLP**) applications. She has a master's degree from Georgia Tech as well, and a bachelor's degree in aerospace engineering from Florida Institute of Technology.

About the reviewer

Valeriy Babushkin is the senior director of data science at X5 Retail Group, where he leads a team of over 80 people in the fields of machine learning, data analysis, computer vision, NLP, R&D, and A/B testing. Valeriy is a Kaggle competition grandmaster and an attending lecturer at the National Research Institute's Higher School of Economics and Central Bank of Kazakhstan.

Valeriy served as a technical reviewer for the books *AI Crash Course* and *Hands-On Reinforcement Learning with Python, Second Edition*, both published by Packt.

Packt is searching for authors like you

If you're interested in becoming an author for Packt, please visit `authors.packtpub.com` and apply today. We have worked with thousands of developers and tech professionals, just like you, to help them share their insight with the global tech community. You can make a general application, apply for a specific hot topic that we are recruiting an author for, or submit your own idea.

Table of Contents

Preface

Part I – Basic Concepts of Discrete Math

1

Key Concepts, Notation, Set Theory, Relations, and Functions

What is discrete mathematics?	4	Definition: Cardinality	12
Elementary set theory	6	Example: Cardinality	12
Definition–Sets and set notation	6	**Functions and relations**	**12**
Definition: Elements of sets	6	Definition: Relations, domains, and ranges	13
Definition: The empty set	7	Definition: Functions	13
Example: Some examples of sets	7	Examples: Relations versus functions	13
Definition: Subsets and supersets	7	Example: Functions in elementary algebra	14
Definition: Set-builder notation	7	Example: Python functions versus mathematical functions	15
Example: Using set-builder notation	8	**Summary**	**16**
Definition: Basic set operations	8		
Definition: Disjoint sets	10		
Example: Even and odd numbers	10		
Theorem: De Morgan's laws	10		
Example: De Morgan's Law	12		

2

Formal Logic and Constructing Mathematical Proofs

Formal Logic and Proofs by Truth Tables	20	Example – all penguins live in South Africa!	23
Basic Terminology for Formal Logic	20	Cores Ideas in Formal Logic	24
Example – an invalid argument	22	Truth Tables	26

Example – The Converse	27
Example – Transitivity Law of Conditional Logic	28
Example – De Morgan's Laws	29
Example – The Contrapositive	30

Direct Mathematical Proofs — 31

Example – Products of Even and Odd Integers	31
Example – roots of even numbers	32
Shortcut – The Contrapositive	33

Proof by Contradiction — 34

Example – is there a smallest positive rational number?	35

Example – Prove $\sqrt{2}$ is an Irrational Number	36
Example – How Many Prime Numbers Are There?	37

Proof by mathematical induction — 38

Example – Adding $1 + 2 + \ldots + n$	39
Example – Space-Filling Shapes	41
Example – exponential versus factorial growth	42

Summary — 44

3

Computing with Base-n Numbers

Understanding base-n numbers — 46

Example – Decimal numbers	46
Definition – Base-n numbers	47

Converting between bases — 47

Converting base-n numbers to decimal numbers	48
Example – Decimal value of a base-6 number	48
Base-n to decimal conversion	48
Example – Decimal to base-2 (binary) conversion	48
Example – Decimal to binary and hexadecimal conversions in Python	50

Binary numbers and their applications — 51

Boolean algebra	52
Example – Netflix users	56

Hexadecimal numbers and their application — 59

Example – Defining locations in computer memory	60
Example – Displaying error messages	62
Example – Media Access Control (MAC) addresses	62
Example – Defining colors on the web	63

Summary — 64

4

Combinatorics Using SciPy

The fundamental counting rule — 66

Definition – the Cartesian product	66

Theorem – the cardinality of Cartesian products of finite sets	66

Definition – the Cartesian product (for n sets)	67	Example – combinations versus permutation for a simple set	71
Theorem – the fundamental counting rule	67	Theorem – combinations of a set	72
Example – bytes	67	Binomial coefficients	72
Example – colors on computers	68	Example – teambuilding	72
		Example – combinations of balls	73

Counting permutations and combinations of objects — 68

Applications to memory allocation — 74

Definition – permutation	68	Example – pre-allocating memory	74
Example – permutations of a simple set	68		
Theorem – permutations of a set	69		

Efficacy of brute-force algorithms — 76

Example – playlists	69	Example – Caesar cipher	76
Growth of factorials	69	Example – the traveling salesman problem	79
Theorem – k-permutations of a set	70		
Definition – combination	71		

Summary — 81

5

Elements of Discrete Probability

The basics of discrete probability — 84

Definition – random experiment	84	Example – tossing many coins	91

Conditional probability and Bayes' theorem — 93

Definitions – outcomes, events, and sample spaces	85	Definition – conditional probability	94
Example – tossing coins	85	Example – temperatures and precipitation	94
Example – tossing multiple coins	85	Theorem – multiplication rules	95
Definition – probability measure	86	Theorem – the Law of Total Probability	96
Theorem – elementary properties of probability	87	Theorem – Bayes' theorem	96
Example – sports	87		

Bayesian spam filtering — 97

Random variables, means, and variance — 98

Theorem – Monotonicity	88	Definition – random variable	99
Theorem – Principle of Inclusion-Exclusion	89	Example – data transfer errors	99
Definition – Laplacian probability	90	Example – empirical random variable	100
Theorem – calculating Laplacian probabilities	90	Definition – expectation	100
Example – tossing multiple coins	91	Example – empirical random variable	101
Definition – independent events	91		

Definition – variance and standard deviation	101	Example – empirical random variable	102
Theorem – practical calculation of variance	102	Google PageRank I	102
		Summary	106

Part II – Implementing Discrete Mathematics in Data and Computer Science

6
Computational Algorithms in Linear Algebra

Understanding linear systems of equations	**110**	Definition – Matrix multiplication	124
Definition – Linear equations in two variables	110	Example – Multiplying matrices by hand and with NumPy	125
Definition – The Cartesian coordinate plane	111	**Solving small linear systems with Gaussian elimination**	**127**
Example – A linear equation	112	Definition – Leading coefficient (pivot)	128
Definition – System of two linear equations in two variables	113	Definition – Reduced row echelon form	128
Definition – Systems of linear equations and their solutions	118	Algorithm – Gaussian elimination	130
Definition – Consistent, inconsistent, and dependent systems	118	Example – 3-by-3 linear system	131
Matrices and matrix representations of linear systems	**119**	**Solving large linear systems with NumPy**	**133**
Definition – Matrices and vectors	119	Example – A 3-by-3 linear system (with NumPy)	133
Definition – Matrix addition and subtraction	121	Example – Inconsistent and dependent systems with NumPy	134
Definition – Scalar multiplication	122	Example – A 10-by-10 linear system (with NumPy)	135
Definition – Transpose of a matrix	123	**Summary**	**137**
Definition – Dot product of vectors	124		

7

Computational Requirements for Algorithms

Computational complexity of algorithms	140	Complexity of common search algorithms	159
Understanding Big-O Notation	145	Linear search algorithm	160
Complexity of algorithms with fundamental control structures	151	Binary search algorithm	161
		Common classes of computational complexity	164
Sequential flow	152	Summary	166
Selection flow	153	References	167
Repetitive flow	155		

8

Storage and Feature Extraction of Graphs, Trees, and Networks

Understanding graphs, trees, and networks	170	Definition: adjacency list	181
		Definition: adjacency matrix	182
Definition: graph	170	Definition: adjacency matrix for a directed graph	184
Definition: degree of a vertex	171		
Definition: paths	172	Efficient storage of adjacency data	186
Definition: cycles	172	Definition: weight matrix of a network	187
Definition: trees or acyclic graphs	173	Definition: weight matrix of a directed network	188
Definition: networks	174		
Definition: directed graphs	175	Feature extraction of graphs	190
Definition: directed networks	176	Degrees of vertices in a graph	190
Definition: adjacent vertices	177	The number of paths between vertices of a specified length	191
Definition: connected graphs and connected components	177		
		Theorem: powers of adjacency matrices	193
Using graphs, trees, and networks	178	Matrix powers in Python	193
Storage of graphs and networks	181	Theorem: minimum-edge paths between v_i and v_j	194
		Summary	195

9
Searching Data Structures and Finding Shortest Paths

Searching Graph and Tree data structures	198	Dijkstra's Algorithm for Finding Shortest Paths	214
Depth-first search (DFS)	199	Dijkstra's algorithm	215
A Python implementation of DFS	201	Applying Dijkstra's Algorithm to a Small Problem	216
The shortest path problem and variations of the problem	205	Python Implementation of Dijkstra's Algorithm	221
Shortest paths on networks	205	Example – shortest paths	225
Beyond Shortest-Distance Paths	206	Example – A network that is not connected	228
Shortest Path Problem Statement	207		
Checking whether Solutions Exist	208	Summary	231
Finding Shortest Paths with Brute Force	212		

Part III – Real-World Applications of Discrete Mathematics

10
Regression Analysis with NumPy and Scikit-Learn

Dataset	236	Least-squares lines with NumPy	245
Best-fit lines and the least-squares method	238	Least-squares curves with NumPy and SciPy	249
Variable	238	Least-squares surfaces with NumPy and SciPy	252
Linear relationship	238		
Regression	238	Summary	255
The line of best fit	240		
The least-squares method and the sum of squared errors	243		

11
Web Searches with PageRank

The Development of Search Engines over time	258	Implementing the PageRank algorithm in Python	268
Google PageRank II	260	Applying the Algorithm to Real Data	273
		Summary	278

12
Principal Component Analysis with Scikit-Learn

Understanding eigenvalues, eigenvectors, and orthogonal bases	280	The scikit-learn implementation of PCA	290
The principal component analysis approach to dimensionality reduction	286	An application to real-world data	294
		Summary	298

Other Books You May Enjoy

Index

Preface

Practical Discrete Mathematics is a comprehensive introduction for those who are new to the mathematics of countable objects. This book will help you get up to speed with using discrete math principles to take your computer science skills to another level. You'll learn the language of discrete mathematics and methods crucial to studying and describing objects and algorithms from computer science and machine learning. Complete with real-world examples, this book covers the internal workings of memory and CPUs, analyzes data for useful patterns, and shows you how to solve problems in network routing, web searching, and data science.

Who this book is for

This book is for computer scientists looking to expand their knowledge of the core of their field. University students seeking to gain expertise in computer science, mathematics, statistics, engineering, and related disciplines will also find this book useful. Knowledge of elementary real-number algebra and basic programming skills in any language are the only requirements.

What this book covers

Part I – Basic Concepts of Discrete Math

Chapter 1, *Key Concepts, Notation, Set Theory, Relations, and Functions,* is an introduction to the basic vocabulary, concepts, and notation of discrete mathematics.

Chapter 2, *Formal Logic and Constructing Mathematical Proofs,* covers formal logic and binary and explains how to prove mathematical results.

Chapter 3, *Computing with Base-n Numbers,* discusses arithmetic in different numbering systems, including hexadecimal and binary.

Chapter 4, *Combinatorics Using SciPy,* explains how to count the elements in certain types of discrete structures.

Chapter 5, Elements of Discrete Probability, covers measuring chance and the basics of Google's PageRank algorithm.

Part II – Implementing Discrete Mathematics in Data and Computer Science

Chapter 6, Computational Algorithms in Linear Algebra, explains how to solve algebra problems with Python using NumPy.

Chapter 7, Computational Requirements for Algorithms, gives you the tools to determine how long algorithms take to run and how much space they require.

Chapter 8, Storage and Feature Extraction of Graphs, Trees, and Networks, covers storing graph structures and finding information about them with code.

Chapter 9, Searching Data Structures and Finding Shortest Paths, explains how to traverse graphs and figure out efficient paths between vertices.

Part III – Real-World Applications of Discrete Mathematics

Chapter 10, Regression Analysis with NumPy, is a discussion on the prediction of variables in datasets containing multiple variables.

Chapter 11, Web Searches with PageRank, shows you how to rank the results of web searches to find the most relevant web pages.

Chapter 12, Principal Component Analysis with Scikit-Learn, explains how to reduce the dimensionality of high-dimensional datasets to save space and speed up machine learning.

To get the most out of this book

Knowledge of elementary real-number algebra and Python SPACE basic programming skills are the main requirements for this book.

You will need to install Python—the latest version, if possible—to run the code in the book. You will also need to install the Python libraries listed in the following table to run some of the code in the book. All code examples have been tested in JupyterLab using a Python 3.8 environment on the Windows 10 OS, but they should work with any version of Python 3 in any OS compatible with it and with any modern integrated development environment, or simply a command line.

Software/hardware covered in the book	OS requirements
Python 3.0 or above	Windows, Mac OS X, or any Linux distribution compatible with Python
Python libraries: NumPy, matplotlib, pandas, scikit-learn, SciPy, seaborn	

More information about installing Python and its libraries can be found in the following links:

- **Python**: https://www.python.org/downloads/
- **matplotlib**: https://matplotlib.org/3.3.3/users/installing.html
- **NumPy**: https://numpy.org/install/
- **pandas**: https://pandas.pydata.org/pandas-docs/stable/getting_started/install.html
- **scikit-learn**: https://scikit-learn.org/stable/install.html
- **SciPy**: https://www.scipy.org/install.html
- **seaborn**: https://seaborn.pydata.org/installing.html

If you are using the digital version of this book, we advise you to type the code yourself or access the code via the GitHub repository (link available in the next section). Doing so will help you avoid any potential errors related to the copying and pasting of code.

Download the example code files

You can download the example code files for this book from GitHub at https://github.com/PacktPublishing/Practical-Discrete-Mathematics. In case there's an update to the code, it will be updated on the existing GitHub repository.

We also have other code bundles from our rich catalog of books and videos available at https://github.com/PacktPublishing/. Check them out!

Download the color images

We also provide a PDF file that has color images of the screenshots/diagrams used in this book. You can download it here: https://static.packt-cdn.com/downloads/9781838983147_ColorImages.pdf.

Conventions used

There are a number of text conventions used throughout this book.

Keywords: indicates keywords and vocabulary.

`Code in text`: Indicates names of scripts, functions, packages, folder names, filenames, file extensions, and pathnames.

A block of code is typeset as follows:

```
import numpy# initialize a matrix
A = numpy.array([[3, 2, 1], [9, 0, 1], [3, 4, 1]])
print(A)
```

The output from code is typeset as follows:

```
[[3 2 1]
 [9 0 1]
 [3 4 1]]
```

Lastly, we have important notes, which appear as follows.

> **Important Note**
> Appear like this.

Get in touch

Feedback from our readers is always welcome.

General feedback: If you have questions about any aspect of this book, mention the book title in the subject of your message and email us at `customercare@packtpub.com`.

Errata: Although we have taken every care to ensure the accuracy of our content, mistakes do happen. If you have found a mistake in this book, we would be grateful if you would report this to us. Please visit `www.packtpub.com/support/errata`, selecting your book, clicking on the Errata Submission Form link, and entering the details.

Piracy: If you come across any illegal copies of our works in any form on the Internet, we would be grateful if you would provide us with the location address or website name. Please contact us at `copyright@packt.com` with a link to the material.

If you are interested in becoming an author: If there is a topic that you have expertise in and you are interested in either writing or contributing to a book, please visit `authors.packtpub.com`.

Reviews

Please leave a review. Once you have read and used this book, why not leave a review on the site that you purchased it from? Potential readers can then see and use your unbiased opinion to make purchase decisions, we at Packt can understand what you think about our products, and our authors can see your feedback on their book. Thank you!

For more information about Packt, please visit `packt.com`.

Part I – Basic Concepts of Discrete Math

Here you will learn the critical vocabulary, notations, and methods of discrete mathematics, including set theory, functions and relations, logic and proofs, arithmetic, counting, and basic probability as applied to computer science.

This part comprises the following chapters:

- *Chapter 1, Key Concepts, Notation, Set Theory, Relations, and Functions*
- *Chapter 2, Formal Logic and Constructing Mathematical Proofs*
- *Chapter 3, Computing with Base-n Numbers*
- *Chapter 4, Combinatorics Using SciPy*
- *Chapter 5, Elements of Discrete Probability*

1
Key Concepts, Notation, Set Theory, Relations, and Functions

This chapter is a general introduction to the main ideas of discrete mathematics. Alongside this, we will go through key terms and concepts in the field. After that, we will cover set theory, the essential notation and notions for referring to collections of mathematical objects and combining or selecting them. We will also think about mapping mathematical objects to one another with functions and relations and visualizing them with graphs.

In this chapter, we will cover the following topics:

- What is discrete mathematics?
- Elementary set theory
- Functions and relations

By the end of the chapter, you should be able to speak in the language of discrete mathematics and understand notation common to the entire field.

> **Important Note**
> Please navigate to the graphic bundle link to refer to the color images for this chapter.

What is discrete mathematics?

Discrete mathematics is the study of countable, distinct, or separate mathematical structures. A good example is a pixel. From phones to computer monitors to televisions, modern screens are made up of millions of tiny dots called pixels lined up in grids. Each pixel lights up with a specified color on command from a device, but only a finite number of colors can be displayed in each pixel.

The millions of colored dots taken together form intricate patterns and give our eyes the impression of shapes with smooth curves, as in the boundary of the following circle:

Figure 1.1 – The boundary of a circle

But if you zoom in and look closely enough, the true "curves" are revealed to be jagged boundaries between differently colored regions of pixels, possibly with some intermediate colors, as shown in the following diagram:

Figure 1.2 – A zoomed-in view of the circle

Some other examples of objects studied in discrete mathematics are logical statements, integers, bits and bytes, graphs, trees, and networks. Like pixels, these too can form intricate patterns that we will try to discover and exploit for various purposes related to computer and data science throughout the course of the book.

In contrast, many areas of mathematics that may be more familiar, such as elementary algebra or calculus, focus on continuums. These are mathematical objects that take values over continuous ranges, such as the set of numbers x between 0 and 1, or mathematical functions plotted as smooth curves. These objects come with their own class of mathematical methods, but are mostly distinct from the methods for discrete problems on which we will focus.

In recent decades, discrete mathematics has been a topic of extensive research due to the advent of computers with high computational capabilities that operate in "discrete" steps and store data in "discrete" bits. This makes it important for us to understand the principles of discrete mathematics as they are useful in understanding the underlying ideas of software development, computer algorithms, programming languages, and cryptography. These computer implementations play a crucial role in applying principles of discrete mathematics to real-world problems.

Some real-world applications of discrete mathematics are as follows:

- **Cryptography**: The art and science of converting data or information into an encoded form that can ideally only be decoded by an authorized entity. This field makes heavy use of number theory, the study of the counting numbers, and algorithms on base-n number systems. We will learn more about these topics in *Chapter 2, Formal Logic and Constructing Mathematical Proofs*.

- **Logistics**: This field makes use of graph theory to simplify complex logistical problems by converting them to graphs. These graphs can further be used to find the best routes for shipping goods and services, and so on. For example, airlines use graph theory to map their global airplane routing and scheduling. We investigate some of these issues in the chapters of *Part II, Implementing Discrete Mathematics in Data and Computer Science*.

- **Machine Learning**: This is the area that seeks to automate statistical and analytical methods so systems can find useful patterns in data, learn, and make decisions with minimal human intervention. This is frequently applied to predictive modeling and web searches, as we will see in *Chapter 5, Elements of Discrete Probability*, and most of the chapters in *Part III, Real-World Applications of Discrete Mathematics*.

- **Analysis of Algorithms**: Any set of instructions to accomplish a task is an algorithm. An effective algorithm must solve the problem, terminate in a useful amount of time, and not take up too much memory. To ensure the second condition, it is often necessary to count the number of operations an algorithm must complete in order to terminate, which can be complex, but can be done through methods of combinatorics. The third condition requires a similar counting of memory usage. We will encounter some of these ideas in *Chapter 4, Combinatorics Using SciPy, Chapter 6, Computational Algorithms in Linear Algebra,* and *Chapter 7, Computational Requirements for Algorithms.*

- **Relational Databases**: They help to connect the different traits between data fields. For example, in a database containing information about accidents in a city, the "relational feature" allows the user to link the location of the accident to the road condition, lighting condition, and other necessary information. A relational database makes use of the concept of set theory in order to group together relevant information. We see some of these ideas in *Chapter 8, Storage and Feature Extraction of Trees, Graphs, and Networks.*

Now that we have a rough idea of what discrete mathematics is and some of its applications, we will discuss set theory, which forms the basis for this field in the next section.

Elementary set theory

"A set is a Many that allows itself to be thought of as a One."

– Georg Cantor

In mathematics, set theory is the study of collections of objects, which is prerequisite knowledge for studying discrete mathematics.

Definition–Sets and set notation

A set is a collection of objects. If a set A is made up of objects $a_1, a_2, \ldots$, we write it as $A = \{a_1, a_2, \ldots\}$.

Definition: Elements of sets

Each object in a set A is called an element of A, and we write $a_n \in A$.

Definition: The empty set

The empty set is denoted $\emptyset$.

Sets may contain many sorts of objects—numbers, points, vectors, functions, or even other sets.

Example: Some examples of sets

Examples of sets include the following:

- The set of prime numbers less than 10 is $A = \{2, 3, 5, 7\}$.
- The set of the three largest cities in the world is *{Tokyo, Delhi, Shanghai}*.
- The natural numbers are a set $N = \{1, 2, 3, ...\}$.
- The integers are a set $Z = \{..., -3, -2, -1, 0, 1, 2, 3, ...\}$.
- If B, C, and D are sets, $A = \{B, C, D\}$ is a set of sets.
- The real numbers are written $R = (-\infty, \infty)$, which consists of the entire number line. Note that it is not possible to list the real numbers within braces, as we can with N or Z.

Definition: Subsets and supersets

A set A is a subset of B if all elements in A are also in B, and we write it as $A \subseteq B$. We call B a superset of A. If A is a subset of B, but they are not the same set, we call A a proper subset of B, and write $A \subset B$.

It is helpful to have an alternative notation in order to construct sets satisfying certain criteria, which we call set-builder notation, defined next.

Definition: Set-builder notation

A set may be written as *{x ∈ A | Conditions}*, which consists of the subset of A such that the given conditions are true.

Sometimes, sets will be expressed as *{x | Conditions}* when it is obvious what kind of mathematical object x is from the context.

Example: Using set-builder notation

Examples of sets constructed by set-builder notation include the following.

- The set of even natural numbers is $\{2, 4, 6, ...\} = \{n \mid n = 2k \text{ for some } k \in N\}$. This is an infinite set where each element n is $2 * k$, where k is some natural number belonging to the set $\{1, 2, 3.....\}$.
- The closed interval of real numbers from a to b is $\{x \in R \mid a \leq x \leq b\} = [a, b]$.
- The open interval of real numbers from a to b is $\{x \in R \mid a < x < b\} = (a, b)$.
- The set $R2 = \{(x, y) \mid x, y \in R\}$ consists of the entire 2D coordinate plane.
- The line with slope 2 and y-intercept 3 is the set $\{(x, y) \in R^2 \mid y = 2x + 3\}$.
- The open ball of radius r and center $(0, 0)$ is $\{(x, y) \in R^2 \mid x^2 + y^2 < r\}$, which is the interior, but not the boundary of a circle.
- A circle of radius r and center $(0, 0)$ is $\{(x, y) \in R2 \mid x2 + y2 = r\}$, which is the boundary of the circle.
- The set of all African nations is $\{x \in \text{Nations} \mid x \text{ is in Africa}\}$.

There are some useful operations that may be done to pairs of sets, which we will see in the next definition.

Definition: Basic set operations

Let A and B be sets. Let's take a look at the basic operations:

- The union of sets A and B is the set of all elements in A or B (or both) and is denoted $A \cup B = \{x \mid x \in A \text{ or } x \in B\}$.
- A union of sets $A_1, A_2, ...$ is denoted $\bigcup_{n=1}^{\infty} A_n$.
- The intersection of sets A and B is the set of all elements in both A and B. It is denoted $A \cap B = \{x \mid x \in A \text{ and } x \in B\}$.
- An intersection of sets $A_1, A_2, ...$ is denoted $\bigcap_{n=1}^{\infty} A_n$.
- The complement of set A is all elements in the set that are not in A and is denoted $A^C = \{x \mid x \notin A\}$.
- The difference between sets A and B is the set of all elements in A, but not B, denoted $A - B = \{x \in A \mid x \notin B\}$.

It is often useful to represent these set operations with Venn diagrams, which are visual displays of sets. Here are some examples of the operations shown previously:

- The following displays $A \cup B$:

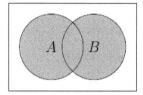

Figure 1.3 – A ⋃ B

- The following displays $A \cap B$:

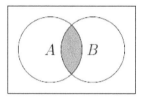

Figure 1.4 – A ⋂ B

- The following displays A^c:

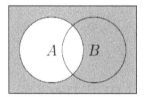

Figure 1.5 – A^c

- The following displays $A - B$:

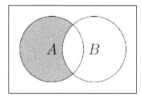

Figure 1.6 – A - B

As an example, consider the following diagram. We can use the language of set theory to describe many aspects of the diagram:

- Elements *a*, *b*, and *d* are in set *A*, which we can write as $a, b, d \in A$.
- Elements *c* and *d* are in set *B*, and $c, d \in B$.
- Element *c* is not in *A*, so we could write $c \notin A$ or $c \in A^C$.
- Element *d* is in both *A* and *B*, or $d \in A \cap B$.
- All four elements are in *A* or *B* (or both), so we could say $a, b, c, d \in A \cup B$:

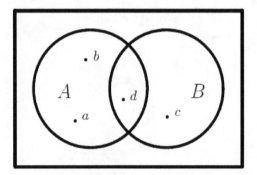

Figure 1.7 – Two sets with some elements

Definition: Disjoint sets

Sets *A* and *B* are disjoint (or mutually exclusive) if $A \cap B = \emptyset$. In other words, the sets share no elements in common.

Example: Even and odd numbers

Consider sets of even natural numbers $E = \{2, 4, 6, ...\}$ and odd natural numbers $O = \{1, 3, 5, ...\}$. These sets are disjoint, $E \cap O = \emptyset$, since no number is both odd and even.

- *E* is a subset of the natural numbers, $E \subseteq N$.
- *O* is a subset of the natural numbers, $O \subseteq N$.

The union of *E* and *O* make up the set of all-natural numbers, $E \cup O = N$.

Theorem: De Morgan's laws

De Morgan's laws state how mathematical concepts are related through their opposites. In set theory, these laws make use of complements to address the intersection and union of sets.

De Morgan's laws can be written as follows:

1. $(A \cup B)^C = A^C \cap B^C$
2. $(A \cap B)^C = A^C \cup B^C$

The following diagrams display De Morgan's laws:

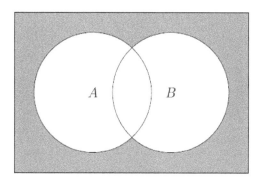

Figure 1.8 – De Morgan's laws $(A \cup B)^C = A^C \cap B^C$

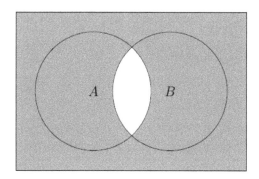

Figure 1.9 – De Morgan's laws $(A \cap B)^C = A^C \cup B^C$

Proof:

Let's now look at the proof of this theorem:

Let $x \in (A \cup B)^C$, then $x \notin (A \cup B)$, which means $x \notin A$ and $x \notin B$, or $x \in A^C$ and $x \in B^C$, or $x \in A^C \cap B^C$. Thus, $(A \cup B)^C$ is a subset of $A^C \cap B^C$.

Next, let $x \in (A^C \cap B^C)$, then $x \in A^C$ and $x \in B^C$, or $x \notin A$ and $x \notin B$, then $x \notin (A \cup B)$ or $x \in (A \cup B)^C$. Like the last step, we see $A^C \cap B^C$ is a subset of $(A \cup B)^C$. Since $(A \cup B)^C$ is a subset of $A^C \cap B^C$ and vice versa, $(A \cup B)^C = A^C \cap B^C$.

The proof of this result is similar and is left as an exercise for the reader.

Notice that the preceding method of proof is designed to show that any element of $(A \cup B)^C$ is an element of $A^C \cap B^C$, and to show that any element of $A^C \cap B^C$ is an element of $(A \cup B)^C$, which establishes that the two sets are the same.

Example: De Morgan's Law

Consider two sets of natural numbers, the even numbers $E = \{2, 4, 6, \ldots\}$ and $A = \{1, 2, 3, 4\}$. If we take the set of elements in either set, or the complement of the union of the sets, we have $(E \cup A)^C = \{1, 2, 3, 4, 6, 8, 10, \ldots\}^C = \{5, 7, 9, \ldots\}$.

De Morgan's law states that the intersection of the complements of the sets should be equal to this. Let's verify that this is true. The complements of the sets are $E^C = \{1, 3, 5, \ldots\}$ and $A^C = \{5, 6, 7, \ldots\}$. The intersection of these complements is $E^C \cap A^C = \{5, 7, 9, \ldots\}$.

Definition: Cardinality

The cardinality, or size, of a set A is the number of elements in the set and is denoted $|A|$.

Example: Cardinality

The cardinalities of some sets are computed here:

- If $A = \{0, 1\}$, then of course its cardinality is $|A| = 2$, since there are two elements in the set.
- The cardinality of the set $B = \{x \in N \mid x < 10\}$ is less obvious, but we can write B more explicitly. It is the set of natural numbers less than 10, so $B = \{1, 2, 3, 4, 5, 6, 7, 8, 9\}$ and, clearly, $|B| = 9$.
- For the set of odd natural numbers, $O = \{1, 3, 5, \ldots\}$, we have an infinite cardinality, $|O| = \infty$, as this sequence goes on forever.

With our knowledge of set theory, we can now move on to learn about relations between different sets and functions, which help us to map each element from a set to exactly one element in another set.

Functions and relations

"Gentlemen, mathematics is a language."

– Josiah Willard Gibbs

We are related to different people in different ways; for example, the relationship between a father and his son, the relationship between a teacher and their students, and the relationship between co-workers, to name just a few. Similarly, relationships exist between different elements in mathematics.

Definition: Relations, domains, and ranges

- A relation r between sets X and Y is a set of ordered pairs (x, y) where $x \in X$ and $y \in Y$.
- The set $\{x \in X \mid (x, y) \in r \text{ for some } y \in Y\}$ is the domain of r.
- The set $\{y \in Y \mid (x, y) \in r \text{ for some } x \in X\}$ is the range of r.

More informally, a relation pairs element of X with one or more elements of Y.

Definition: Functions

- A function f from X to Y, denoted $f: X \to Y$, is a relation that maps each element of X to exactly one element of Y.
- X is the domain of f.
- Elements of the function (x, y) are sometimes written $(x, f(x))$.

As the definitions reveal, functions are relations, but must satisfy a number of additional assumptions, in other words, every element of X is mapped to exactly 1 element of Y.

Examples: Relations versus functions

Let's look at $X = \{1, 2, 3, 4, 5\}$ and $Y = \{2, 4, 6, \ldots\}$. Consider two relations between X and Y:

- $r = \{(3, 2), (3, 6), (5, 6)\}$
- $s = \{(1, 4), (2, 4), (3, 8), (4, 6), (5, 2)\}$

The domain of r is $\{3,5\}$ and the range of r is $\{2, 6\}$ while the domain of s is all of X and the range of s is $\{2, 4, 6, 8\}$.

Relation r is not a function because it maps 3 to both 2 and 6. However, s is a function with domain X since it maps each element of X to exactly one element of Y.

Example: Functions in elementary algebra

Elementary algebra courses tend to focus on specific sorts of functions where the domain and range are intervals of the real number line. Domain values are usually denoted by x and values in the range are denoted by y because the set of ordered pairs (x, y) that satisfy the equation $y = f(x)$ plotted on the Cartesian xy-plane form the graph of the function, as can be seen in the following diagram:

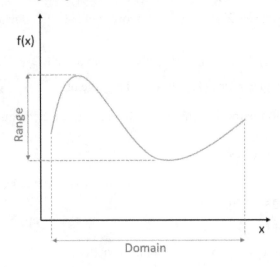

Figure 1.10 – Cartesian xy-plane

While this typical type of functions may be familiar to most readers, the concept of a function is more general than this. First, the input or the output is required to be a number. The domain of a function could consist of any set, so the members of the set may be points in space, graphs, matrices, arrays or strings, or any other types of elements.

In Python and most other programming languages, there are blocks of code known as "functions," which programmers give names and will run when you call them. These Python functions may or may not take inputs (referred to as "parameters") and return outputs, and each set of input parameters may or may not always return the same output. As such, it is important to note Python functions are not necessarily functions in the mathematical sense, although some of them are.

This is an example of conflicting vocabulary in the fields of mathematics and computer science. The next example will discuss some Python functions that are, and some that are not, functions in the mathematical sense.

Example: Python functions versus mathematical functions

Consider the `sort()` Python function, which is used for sorting lists. See this function applied to two lists – one list of numbers and one list of names:

```
numbers = [3, 1, 4, 12, 8, 5, 2, 9]
names = ['Wyatt', 'Brandon', 'Kumar', 'Eugene', 'Elise']

# Apply the sort() function to the lists
numbers.sort()
names.sort()

# Display the output
print(numbers)
print(names)
```

The output is as follows:

```
[1, 2, 3, 4, 5, 8, 9, 12]
['Brandon', 'Elise', 'Eugene', 'Kumar', 'Wyatt']
```

In each case, the `sort()` function sorts the list in ascending order by default (with respect to numerical order or alphabetical order).

Furthermore, we can say that `sort()` applies to any lists and is a function in the mathematical sense. Indeed, it meets all the criteria:

1. The domain is all lists that can be sorted.
2. The range is the set of all such lists that have been sorted.
3. `sort()` always maps each list that can be inputted to a unique sorted list in the range.

Consider now the Python function random, `shuffle()`, which takes a list as an input and puts it into a random order. (Just like the shuffle option on your favorite music app!) Refer to the following code:

```
import random

# Set a random seed so the code is reproducible
random.seed(1)
```

```
# Run the random.shuffle() function 5 times and display the
  # outputs
for i in range(0,5):
  numbers = [3, 1, 4, 12, 8, 5, 2, 9]
  random.shuffle(numbers)
  print(numbers)
```

The output is as follows:

```
[8, 4, 9, 2, 1, 3, 5, 12]
[5, 1, 3, 8, 2, 12, 9, 4]
[2, 1, 12, 9, 5, 4, 8, 3]
[1, 2, 3, 12, 5, 8, 4, 9]
[5, 8, 9, 12, 4, 3, 2, 1]
```

This code runs a loop where each iteration sets the list numbers to [3, 1, 4, 12, 8, 5, 2, 9], applies the shuffle function to it, and prints the output.

In each iteration, the Python function shuffle() takes the same input, but the output is different each time. Therefore, the Python function shuffle() is not a mathematical function. It is, however, a relation that can pair each list with any ordering of itself.

Summary

In this chapter, we have discussed the meaning of discrete mathematics and discrete objects. Furthermore, we provided an overview of some of the many applications of discrete mathematics in the real world, especially in the computer and data sciences, which we will discuss in depth in later chapters.

In addition, we have established some common language and notation of importance for discrete mathematics in the form of set notation, which will allow us to refer to mathematical objects with ease, count the size of sets, represent them as Venn diagrams, and much more. Beyond this, we learned about a number of operations that allow us to manipulate sets by combining them, intersecting them, and finding complements. These give rise to some of the foundational results in set theory in De Morgan's laws, which we will make use of in later chapters.

Lastly, we took a look at the ideas of functions and relations, which map mathematical objects such as numbers to one another. While certain types of functions may be familiar to the reader from high school or secondary school, these familiar functions are typically defined on continuous domains. Since we focus on discrete, rather than continuous, sets in discrete mathematics, we drew the distinction between the familiar idea and a new one we need in this field. Similarly, we showed the difference between functions in mathematics and functions in Python and saw that some Python "functions" are mathematical functions, but others are not.

In the remaining four chapters of *Part I: Core Concepts of Discrete Mathematics*, we will fill our discrete mathematics toolbox with more tools, including logic in *Chapter 2, Formal Logic and Constructing Mathematical Proofs*, numerical systems, such as binary and decimal, in *Chapter 3, Computing with Base n Numbers*, counting complex sorts of objects, including permutations and combinations, in *Chapter 4, Combinatorics Using SciPy*, and dealing with uncertainty and randomness in *Chapter 5, Elements of Discrete Probability*. With this array of tools, we will be able to consider more and more real-world applications of discrete mathematics.

2
Formal Logic and Constructing Mathematical Proofs

This chapter is an introduction to formal logic and mathematical proofs. We'll first introduce some primary results of formal logic and prove logical statements with the use of truth tables. In the remainder of the chapter, we'll consider the most common methods of mathematical proofs (direct proof, proof by contradiction, and proof by mathematical induction) to build skills that you will need for more complex problems to come later.

In this chapter, we will cover the following topics:

- Formal logic and proofs by truth tables
- Direct mathematical proofs
- Proof by contradiction
- Proof by mathematical induction

By the end of the chapter, you will have a grasp of how formal logic provides a grounding for deductive thought, you will have learned how to model logical problems with truth tables, you will have proved claims with truth tables, and you will have learned how to construct mathematical proofs using several methods: direct proof, proof by contradiction, and proof by mathematical induction. This short introduction to mathematical proofs will help you to learn how to think like a mathematician, use powerful deductive thought, and learn the later material in the book.

> **Important Note**
> Please navigate to the graphic bundle link to refer to the color images for this chapter.

Formal Logic and Proofs by Truth Tables

We will be interested in arguments about mathematical structures and mathematical proofs throughout the book so that we can establish mathematical truths that will be used in practical problems. For this reason, in this section, we wish to establish some familiarity with the strict logic required to establish some mathematical theory that allows us to solve practical mathematical problems.

The foundation of all mathematics is logic, which studies how we can construct logically sound arguments that show that certain assumptions lead to certain conclusions with no doubt. In particular, **formal logic** abstracts away any specifics of the particular arguments being constructed in order to focus on the structure of the arguments, which can establish some general principles or shortcuts that can be used in specific arguments. Aristotle developed many principles of syllogistic logic, which is logic focusing on arguments that deductively lead from some assumptions to a conclusion. This work, dating all the way back to the 300s BCE, in fact, is still widely used today. The modern study of formal logic uses and builds upon the pioneering work of Aristotle.

Basic Terminology for Formal Logic

Before proceeding to study formal logic, we need to define some terms and notation to facilitate the discussion. Informally, logic studies how some statements lead to certain consequences. This sounds abstract, so let's consider an example.

Suppose we want to use some simple mathematical reasoning to show that if a positive integer is a multiple of 4, then it is also a multiple of 2. Of course, we probably all intuitively know that this is true based on our experience with arithmetic, but let's carefully write down some reasoning for this claim, step by step, as an example:

1. n is a positive integer.
2. n is a multiple of 4.
3. There exists some positive integer, m, where $n = 4m$.
4. If we factor out 2 on the right side of the equation, we find $n = 2(2m)$.
5. Therefore, n is a multiple of 2.

Let's break this down into pieces and define them in the context of the vocabulary of formal logic:

- Each line of a chain of reasoning that is either true or false is called a statement:

    a) All five lines in the preceding reasoning are statements.

- A collection of statements is called an **argument**:

    a) The whole collection of statements 1–5 makes up an argument.

    b) Note that the word "argument" may be used differently in everyday conversation, but the arguments studied by formal logic must not include any ambiguity, only statements.

- Exactly one statement of an argument is called the **conclusion**:

    a) Statement 5 is the conclusion.

    b) Conclusions usually come at the end.

    c) Conclusions are usually things we would like to prove in mathematical arguments.

- All other statements of the argument are called **premises**:

    a) Statements 1–4 are premises.

- An argument is called **valid** if the conclusion must be true when all the premises are true:

    a) The preceding argument is valid because statement 5 (the conclusion) *must be* true when the first four statements (or premises) are true.

- Any argument that is not valid is called **invalid**.

In other words, in a valid argument, the premises must unambiguously lead to the conclusion, as is true in our preceding simple mathematical argument.

An invalid argument is one where all the premises could be true, but the conclusion is still false. To make this clearer, let's consider an example.

Example – an invalid argument

Consider the following argument:

1. n is a positive integer.
2. n is a multiple of 3.
3. n is a multiple of 5.
4. 3 and 5 are both odd numbers.
5. Therefore, n is an odd number.

So, we have a positive integer, n, which is a multiple of both 3 and 5, which are odd numbers. Assume statements 1–4 are true premises and statement 5 is the conclusion of the argument. Is this a valid argument?

It makes some sense; lots of multiples of 3 are odd:

$$3, 9, 15, \ldots$$

And lots of multiples of 5 are odd:

$$5, 15, 25, \ldots$$

So, does it make sense to conclude that n is odd? No! There are some numbers that are multiples of both 3 and 5 that are not odd, as follows:

$$30, 60, 90, \ldots$$

These are even numbers, so statements 1–4 could be true and n could still be an even number—that is, statement 5 is false. In other words, it is possible for all premises of the argument (statements 1–4) to be true but for the conclusion of the argument (statement 5) to be false simultaneously, so this argument is invalid.

A point that might be surprising is that a valid argument is not always a good argument practically speaking. Let's consider an example.

Example – all penguins live in South Africa!

Consider the following silly argument:

1. All penguins are orange.
2. All orange animals live in countries on the equator.
3. The equator passes through only one country.
4. South Africa is on the equator.
5. All penguins live in South Africa.

Suppose statements 1 and 2 are true. Therefore, all penguins are orange and all orange animals live in countries on the equator, which means all penguins live in countries on the equator. If statements 3 and 4 are true, the only country the equator passes through is South Africa. Combining these two, we can conclude that the only country penguins could live in is South Africa. Thus, these statements imply statement 5 is true.

Here, if the premises (statements 1–4) are true, then the conclusion (statement 5) must be true. Thus, the argument is valid by definition.

There is clearly a problem: none of these premises are actually true! Penguins are not ordinarily orange, many penguins live in cold climates far from the equator, orange animals such as tigers live in countries that are not on the equator, the equator passes through many countries, and South Africa is not one of those countries:

Figure 2.1 – Penguins are certainly not all orange (left) and the equator (the dotted line on the map) is nowhere near South Africa (right)!

As we can see, a valid argument is not always a "good" argument, practically speaking. It simply means that *if* the premises are true, *then* the conclusion is true. There is no requirement for the premises to *actually* be true.

This may seem unusual, but it reveals something important: logic studies the consequences of assumptions we choose to make. It is not necessarily concerned with what is true, *except* for the matter of whether the premises imply the conclusions in an argument.

Next, let's introduce some common notations for writing about arguments to facilitate some analysis of arguments we would like to do.

Cores Ideas in Formal Logic

We will represent statements (also frequently called **propositions** in this context) with single lowercase letters, typically p, q, and r.

In logical arguments, we frequently want to modify propositions and combine propositions to build compound propositions that are more complex or more interesting. For the upcoming ideas, let's consider two simple propositions about a positive integer, n:

- p: 5 is a multiple of 2
- q: 6 is a multiple of 3

For example, we might want to form a proposition "p and q" or, in more readable terms, "5 is a multiple of 2 and 6 is a multiple of 3," which is still a proposition, just a more complex proposition. As a proposition, of course, it still may be true or false.

More formally, **logical connectives** are words or symbols that connect or modify propositions. There is some common notation used for many common connectives. Some of the most common are defined here and the verbal equivalent for the preceding example is given for each:

- The **negation** of a proposition is denoted $\sim p$, which is true only when p is *not* true.

    a) "5 is not a multiple of 2," which is true since p is false.
- The **conjunction** of two propositions is true only when both p *and* q are true and it is written as follows:

$$p \wedge q$$

    a) "5 is a multiple of 2 and 6 is a multiple of 3," which is false since p is false.

- The **disjunction** of two propositions is true when p or q (or both) is true, and it is written as follows:
$$p \vee q$$
    a) "5 is a multiple of 2 or 6 is a multiple of 3," which is true since q is true.

    b) The disjunction is sometimes called the *inclusive* "or."

- The **conditional** or **implication** is true if p is false or q is true and is written as follows:
$$p \to q$$
    a) "If 5 is a multiple of 2, then 6 is a multiple of 3," which is true since p is false.

    b) You can think of a conditional as saying that q is a consequence of p being true. It does not say anything about q if p is false.

    c) Stated in another way, a conditional is only false in the situation where p is true and q is false. In other words, if the conditional is true, p cannot be true unless q is also true.

    d) We will frequently say p implies q.

- The **biconditional** is true if p and q are both true or both false and is written as follows:
$$p \leftrightarrow q$$
    a) "5 is a multiple of 2 if and only if 6 is a multiple of 3," which is false since p is false, but q is true.

    b) Stated a different way, this means p and q are equivalent propositions.

The following figure shows a summary of the common logical connectives we have discussed:

Connective	Notation	English Equivalent
Negation	$\sim p$	not p
Conjunction	$p \wedge q$	p and q
Disjunction	$p \vee q$	p or q
Conditional	$p \to q$	if p, then q
Biconditional	$p \leftrightarrow q$	p if and only if q

Figure 2.2 – Logical connectives

Next, we will learn about truth tables, which provide us with a way to determine whether different compound propositions are equivalent or whether they disagree with one another under some circumstances.

Truth Tables

As you might suspect, it is possible to build complex propositions by combining more and more simple propositions with logical connectives, as follows:

$$(p \to q) \land (q \to r) \to (p \to r)$$

It would be somewhat difficult to determine whether this is a valid argument by pure thought, so a diagram would be helpful. This is exactly what **truth tables** do. They let us break complex logical propositions down into their component parts and determine whether arguments are valid.

More specifically, a truth table is a table of binary values (0 for false and 1 for true), where we consider every possible combination of truth-values (true or false) of the simple propositions and can determine truth-values by applying one logical connective at a time. As an exercise, let's build a truth table for each of the common logical connectives. The first one is the negation:

p	$\sim p$
0	1
1	0

Figure 2.3 – Truth table for the negation connective

This truth table is small—it only involves one proposition, which can be true or false. Of course, the negation just has opposite truth-values in each case.

The other logical connectives involve two propositions, so there are more states—we need to consider every combination of truth-values for each proposition. We present them in the following figure:

Conjunction			Disjunction			Conditional			Biconditional		
p	q	$p \land q$	p	q	$p \lor q$	p	q	$p \to q$	p	q	$p \leftrightarrow q$
0	0	0	0	0	0	0	0	1	0	0	1
0	1	0	0	1	1	0	1	1	0	1	0
1	0	0	1	0	1	1	0	0	1	0	0
1	1	1	1	1	1	1	1	1	1	1	1

Figure 2.4 – Truth tables for the binary logical connectives

These tables are pretty simple to create for these simple logical connectives, and more or less simply represent the definitions in a table form.

Let's see how we can check whether some arguments are equivalent with some examples.

Example – The Converse

A conditional, "if p, then q" looks as follows:

$$p \to q$$

The **converse** is the conditional in the opposite direction, "if q, then p," which looks as follows:

$$q \to p$$

The question is: are these propositions equivalent to each other? Let's construct a truth table containing both of these propositions and see whether they are equivalent:

p	q	$p \to q$	$q \to p$	$(p \to q) \leftrightarrow (q \to p)$
0	0	1	1	1
0	1	1	0	0
1	0	0	1	0
1	1	1	1	1

Figure 2.5 – Truth table for a conditional and its converse. The columns containing the truth-values of the two propositions being compared are shaded

Notice that the two conditionals do not always agree with one another. If one statement is true but the other is false, the conditionals do not have the same truth-values, so they are not equivalent. In other words, the biconditional is false:

$$(p \to q) \leftrightarrow (q \to p)$$

This means that, if p implies q, it is not necessarily true that q implies p. This should make some sense intuitively. For example, it is true that "if n is divisible by 4, then n is divisible by 2," but it is not true that "if n is divisible by 2, then n is divisible by 4" because n could be 6 or 10, which are not divisible by 4.

Let's consider another example that has some important consequences for arguments.

Example – Transitivity Law of Conditional Logic

If we show p implies q and q implies r, it seems intuitive that we could simply say p implies r, but can we show this with a truth table? In other words, can we establish the difficult proposition we wrote previously?

$$(p \to q) \land (q \to r) \to (p \to r)$$

Let's create a truth table. This time, we have three basic propositions in p, q, and r, so our truth table will need to consider every combination of truth-values of these three propositions:

p	q	r	$p \to q$	$q \to r$	$(p \to q) \land (q \to r)$	$p \to r$	$(p \to q) \land (q \to r) \to (p \to r)$
0	0	0	1	1	1	1	1
0	0	1	1	1	1	1	1
0	1	0	1	0	0	1	1
0	1	1	1	1	1	1	1
1	0	0	0	1	0	0	1
1	0	1	0	1	0	1	1
1	1	0	1	0	0	0	1
1	1	1	1	1	1	1	1

Figure 2.6 – A truth table confirming the transitive rule. The columns containing the truth-values of the two propositions being compared are shaded

Therefore, we see the proposition in the rightmost column is always true. This means, regardless of the truth-values of the propositions p, q, and r, the proposition is true. Therefore, anytime we can prove p implies q and q implies r, we have automatically proven p implies r. As a result, we can chain implications in a sequence to establish a conclusion. This is sometimes called the **transitivity law** for implications.

Let's try another example.

Example – De Morgan's Laws

Suppose we have two propositions, p and q, and consider the negation of their conjunction. We would like to prove De Morgan's laws, one of which states this is equivalent to the disjunction of their negations. In symbols, this looks as follows:

$$\sim(p \land q) \leftrightarrow (\sim p \lor \sim q)$$

In simpler words, this says that "p and q are not both true" is equivalent to stating "p is not true, or q is not true." Let's use a truth table to see whether it is true:

p	q	$p \land q$	$\sim(p \land q)$	$\sim p$	$\sim q$	$\sim p \lor \sim q$	$\sim(p \land q) \leftrightarrow (\sim p \lor \sim q)$
0	0	0	1	1	1	1	1
0	1	0	1	1	0	1	1
1	0	0	1	0	1	1	1
1	1	1	0	0	0	0	1

Figure 2.7 – A truth table confirming one of De Morgan's laws. The columns containing the truth-values of the two propositions being compared are shaded

As we see from the truth table, this one of De Morgan's laws is true since the two sides of the biconditional are equivalent to one another. It should be noted there is another of De Morgan's laws that may be written as follows:

$$\sim(p \lor q) \leftrightarrow (\sim p \land \sim q)$$

This one says that "p or q is not true" (keeping in mind this is the *inclusive* "or") is equivalent to "p is not true and q is not true." This one can be proven very similarly in the next truth table:

p	q	$p \lor q$	$\sim(p \lor q)$	$\sim p$	$\sim q$	$\sim p \land \sim q$	$\sim(p \lor q) \leftrightarrow (\sim p \land \sim q)$
0	0	0	1	1	1	1	1
0	1	1	0	1	0	0	1
1	0	1	0	0	1	0	1
1	1	1	0	0	0	0	1

Figure 2.8 – A truth table confirming another of De Morgan's laws. The columns containing the truth-values of the two propositions being compared are shaded

30   Formal Logic and Constructing Mathematical Proofs

These laws are named for Augustus De Morgan, who first stated them in terms of formal logic as we have here in the 1800s, although the ideas were known before this. These laws can allow us to do a trick to convert something we would like to prove, such as "p and q are not both true," and instead prove "p is false or q is false," which may sound like an obvious step, but when the propositions become much more complex, it is not always so easy.

Let's look at one more example that can provide a helpful trick for proofs that is far less obvious.

Example – The Contrapositive

Suppose we need to prove p implies q:

$$p \to q$$

But this proves difficult. It turns out that there is an alternative: the **contrapositive**. The contrapositive says, "not q implies not p," which looks as follows:

$$\sim q \to \sim p$$

The contrapositive seems similar to the converse, except for the negations on each side of the conditional. Let's compare them on a truth table:

p	q	$p \to q$	$\sim p$	$\sim q$	$\sim q \to \sim p$	$(p \to q) \leftrightarrow (\sim q \to \sim p)$
0	0	1	1	1	1	1
0	1	1	1	0	1	1
1	0	0	0	1	0	1
1	1	1	0	0	1	1

Figure 2.9 – A truth table confirming the conditional is equivalent to the contrapositive. The columns containing the truth-values of the two propositions being compared are shaded

As we can see from the truth table, the conditional is true precisely when the contrapositive is true, so the two ideas are equivalent. We will see a way in which this can be used to drastically shorten a mathematical proof as follows.

With these formal logical results in hand, we will move on to discussing some different styles of mathematical proofs, all of which are valid logical arguments, and some examples where they can be applied. We encourage you to read the mathematical claim we make and try to prove it on your own before reading our proofs. The best way to improve with making mathematical arguments is to try them on your own.

The subject matter is not especially important to building skills with mathematical proofs, so we have chosen many examples using only simple topics such as whole numbers. We believe this is the best way to learn how to construct proofs so that you can focus on the structure of the arguments without too much distraction. This will allow you to focus on the skills needed to practice the careful styles of thought required to establish mathematical truth.

In this section, we learned about the basic terminology for formal logic, common logical connectives such as negation, conjunction, disjunction, conditional, and so on. We also learned about truth tables and came up with multiple truth tables for examining the transitivity law of conditional logic, contrapositive, De Morgan's laws, and so on.

In the next section, we will use the ideas covered in this section and investigate mathematical proofs.

Direct Mathematical Proofs

In this section, we will look into how mathematical proofs are constructed and understand this with a few simple examples.

The simplest way to establish a mathematical truth is through a direct proof that shows the definitions of the terms led through a sequence of deductions that lead to the conclusion we wish to prove.

Let's look at a simple example and construct our own proof showing that the product of an even and an odd integer is itself an even number.

Example – Products of Even and Odd Integers

Let x be an even integer. This means x is a multiple of 2, so there exists an integer n where we have the following:

$$x = 2n$$

Let y be an odd integer. This means y is not a multiple of 2, which means when we divide it by 2, we will have a remainder of 1, which means there is an integer m such that we have the following:

$$y = 2m + 1$$

If we multiply them together, we find the following:

$$xy = (2n)(2m + 1)$$

$$xy = 4nm + 2n$$

$$xy = 2(2nm + n)$$

Since $2nm + n$ is made up of a product and sum of integers, it is also an integer. Therefore, the product of x and y equals 2 multiplied by an integer. This means the product xy is a multiple of 2—in other words, it is an even integer.

As you can see, the ideas here were simple. We wrote down precisely what it means for x to be even and y to be odd. Then, we did some algebraic manipulations and found that the product xy is a multiple of 2 and, therefore, an even number.

The structure of any mathematical proof has some things in common: each step leads logically to the next. Constructing a valid proof, however, must follow very strict deductive steps. There can be absolutely no guesswork in a mathematical proof. We can add no extra assumptions without changing the statement we have proven. This can make proofs difficult to construct sometimes and the conclusions of a single proof are often somewhat narrow, but mathematical proofs establish quite possibly the closest thing to absolute truth humans can produce.

Let's try another proof, again regarding even and odd integers.

Example – roots of even numbers

Suppose n^2 is an even number where n is a positive integer. We will determine whether n is even or odd. There are only two possibilities: n is even or n is odd. If n is even, there is a non-negative integer k where we have the following:

$$n = 2k$$

In this case, if we square n, we have the following:

$$n^2 = (2k)(2k)$$

$$n^2 = 4k^2$$

$$n^2 = 2(2k^2)$$

Therefore, n^2 is 2 times $2k^2$. Clearly, $2k^2$ is an integer, so this means n^2 is an even number.

We have not proven what we wanted to prove yet. We have instead proven the converse of the goal: if n is even, then n^2 is even. As we saw in the previous section, the converse is not equivalent to the original conditional proposition. We are not finished, then. But, if we can show the only other possibility, that n is odd implies n^2 is odd, then we can know that n must be even.

If n is odd, there is a non-negative integer m where we have the following:

$$n = 2m + 1$$

In this case, if we square n, we have the following:

$$n^2 = (2m + 1)^2$$
$$n^2 = (2m + 1)(2m + 1)$$
$$n^2 = 4m^2 + 2m + 2m + 1$$
$$n^2 = 4m^2 + 4m + 1$$
$$n^2 = 2(2m^2 + 2m) + 1$$

Therefore, n^2 equals 2 times $2m^2 + 2m$ plus 1. Clearly $2m^2 + 2m$ is an integer, so this means n^2 is an odd number under the assumption we made, that n was odd.

In summary, we showed if n is odd, n^2 must be odd, and if n is even, n^2 must be even. Since these are the only possible states for n, this means if n^2 is even, then n must be even.

Shortcut – The Contrapositive

We have successfully proven the preceding statement, but note that we can break the argument up into separate pieces:

$$p: n^2 \text{ is even}$$
$$q: n \text{ is even}$$

We also proved that p implies q. However, it took significant time and effort and really did not flow from start to finish so easily. As we showed in the previous section, proving the conditional p implies q is equivalent to the contrapositive:

$$\sim q \rightarrow \sim p$$

In this problem, this implication would read "if n is not even, then n^2 is not even," or, we might say "if n is odd, then n^2 is odd." We actually proved this in the example using the second batch of equations. It turns out, proving just that is the contrapositive, which is equivalent to the original goal of proving p implies q, so the contrapositive was a more efficient approach.

In this section, we learned about some simple mathematical proofs and worked on constructing them. We recommend adding the contrapositive to your toolbox of proof techniques for this sort of situation. It is helpful to try to use it when a direct proof seems difficult, but it is easy to state the negations of the propositions making up the conditional we hope to prove.

In the next section, we will learn about proving by contradiction.

Proof by Contradiction

In this section, we will learn about using contradiction for mathematical proofs. Proof by contradiction is a method of proof where you first assume the claim you wish to prove is false, and then prove through a series of logical deductions that this assumption results in a contradictory claim. If this happens, and we have made no errors, this assumption that the claim was false must have been incorrect. Thus, the claim must be true.

While this idea may make sense abstractly and we see the proof method is confirmed by formal logic, the authors believe the method is best demonstrated by examples if you hope to build some intuitive understanding of the approach, learn when it is likely to be effective, and construct your own mathematical proofs.

First, let's review some ideas we all probably learned in primary school. Recall a real number x is called *rational* if it can be written as a *ratio*:

$$x = \frac{a}{b}$$

Here, a and $b \neq 0$ are relatively prime integers—that is, numbers who share no common factors or only a common factor of 1—a and b may be negative or positive whole numbers, and a could be 0. For example, the following numbers are rational:

$$0.5, \frac{5}{7}, \frac{10}{100}, \frac{40}{2}, \frac{-1291034}{12812008}, 3, \frac{\pi}{2\pi}, 0$$

Of these rational numbers, note that only one of these numbers, five-sevenths, is actually written as a ratio of two relatively prime numbers. The keywords in the definition of rational numbers are that they *can be* written in this way. Note that all the numbers listed can be written this way, meaning they are in fact rational numbers:

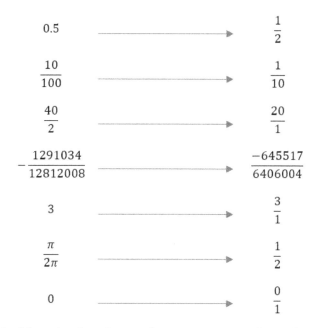

Figure 2.10 – Each of the rational numbers can be written as a ratio of two relatively prime integers

It would not be unexpected if your first question is "Are there any numbers that cannot be written as a fraction?" The answer is certainly yes, but the great ancient Greek mathematicians such as Pythagoras and Euclid debated this question for centuries before it was settled that, in fact, there are numbers that cannot be written as such a ratio. So, this is a good question, and you are in good company if you thought to ask it!

Let's see a couple of examples related to rational numbers.

Example – is there a smallest positive rational number?

The problem here is simple to state: is there a smallest positive rational number? But how can we tackle this? It seems unlikely that we could create a tiny number and somehow claim it is the smallest possible one, although it is also not clear that we can say there isn't such a number. Since no direct path to a proof seems obvious, let's try to prove there is no such number by proof by contradiction.

Suppose x is the smallest positive rational number. Since it is rational, we can write the following:

$$x = \frac{a}{b}$$

Here, a and b are relatively prime integers, both of which have the same sign since x is positive. If we divide x by 2, we get a smaller number:

$$y = \frac{a}{2b}$$

This number is still positive as no signs have changed. b is a nonzero integer, so $2b$ is also a nonzero integer of the same sign. Therefore, y is a positive rational number that is less than x. This contradicts the assumption that x is the smallest integer. Thus, if you give me any rational number, I can always give you a smaller one by dividing it by 2, so there is no smallest positive rational number.

This was a nice, simple example of proof by contradiction, but let's try another one that is pretty simple in principle, but probably not at all obvious.

Example – Prove $\sqrt{2}$ is an Irrational Number

In this example, we will prove the square root of 2 is irrational, which should put this question to rest. In other words, we will prove the square root of 2 is not rational. We will set up a proof by contradiction.

First, assume the square root of 2 is rational. Therefore, by definition, there exist relatively prime numbers a and b where we have the following:

$$\sqrt{2} = \frac{a}{b}$$

But if we square both sides of the equation, we find the following:

$$2 = \frac{a^2}{b^2}$$

$$2b^2 = a^2$$

Since b is an integer, so is b^2, so a^2 is two times an integer, which means $a2$ is a multiple of 2—in other words, a^2 is an even number. As we have proven previously, this means a must be an even number, so there is an integer n where $a = 2^n$, so we can rewrite the preceding equation as follows:

$$2b^2 = (2n)^2$$
$$2b^2 = 4n^2$$
$$b^2 = 2n^2$$

Therefore, b^2 is an even number, which we have shown implies b is an even number.

We have shown both a and b are even numbers, so they share a factor of 2, meaning they are not relatively prime integers. We previously assumed the square root of 2 was rational and could be written as the ratio of relatively prime integers a and b. Then, the assumption that the square root of 2 is irrational leads to a contradiction that a and b both are and are not relatively prime integers.

Next is a famous example of proof by contradiction regarding prime numbers used by Euclid in approximately 300 BCE. It is actually one of the first known uses of proof by contradiction.

Example – How Many Prime Numbers Are There?

A prime number is a positive integer greater than 1 that is only divisible by 1 and itself. The first few prime numbers are 2, 3, 5, 7, and 11. Note that the numbers we skipped have divisors other than 1 and the number itself. Clearly, 4, 6, 8, and 10 are divisible by 2 and, indeed, all even numbers except 2 will be prime. 9 is an odd number, but it is not prime since it is divisible by 3.

The prime numbers are sometimes called the building blocks of the positive integers because all positive integers can be written as a product of a unique set of prime numbers, called its prime factorization. Take the following example:

$$15 = 3 \cdot 5$$
$$108 = 2 \cdot 2 \cdot 3 \cdot 3 \cdot 3$$

Indeed, no matter how large the initial number, this can be done! Another example is the following:

$$35609874300 = 2 \cdot 2 \cdot 3 \cdot 3 \cdot 5 \cdot 5 \cdot 7 \cdot 11 \cdot 13 \cdot 29 \cdot 29 \cdot 47$$

Once again, this is made up entirely of prime factors. In each case, the numbers cannot be broken down into smaller factors, so these factorizations are unique.

The result is now called the fundamental theorem of arithmetic. But interest in primes goes back to at least ancient Egypt. The Rhind Mathematical Papyrus is an Egyptian artifact dating to 1500 BCE with some computations with primes! But we know much more about the work of ancient Greeks mathematicians with primes. They were quite intrigued by prime numbers. In fact, Euclid proved prime factorizations are unique for all numbers in approximately 300 BCE. A question that arose millennia ago was: "How many prime numbers are there?" According to Euclid, there are infinitely many. But how could he know that? Dealing with infinity can be subtle, so it seems impractical to attempt to prove this directly. In such a situation, where a direct path to a proof seems difficult, proof by contradiction is one of the tools in the toolbox of a seasoned mathematician. Let's try it!

Assume there are finitely many prime numbers. Without loss of generality, suppose the number of primes is a finite positive integer m and let's name all primes $p_1, p_2, ..., p_m$. Let n be a number equal to the product of all the primes plus 1:

$$n = p_1 p_2 \cdots p_m + 1$$

This means $n - 1$ is divisible by $p_1, p_2, ..., p_m$—that is, all of the primes. Each prime number is greater than 1 by definition. Therefore, dividing n by any prime number would have a remainder of 1. Therefore, n is not divisible by any of these prime numbers.

By Euclid's fundamental theorem of arithmetic, all positive integers have a unique prime factorization, so there must be another prime number not in our set. This contradicts the assumption that there are m primes, which was an arbitrary choice of number, so the assumption that there are finitely many primes leads to this contradiction. Hence, the opposite must be true: there are infinitely many primes.

Now, this has completed the proof, but let's zoom in on one point. We said m was an arbitrary choice, which led to the conclusion we made previously. However, this point is not too obvious to the uninitiated.

Let's think about this. If we assumed there were m primes, we concluded there are at least $m + 1$ prime numbers. Say we repeat this argument with the following:

$$n = p_1 p_2 \cdots p_m p_{m+1} + 1$$

We will conclude there are at least $m + 2$ primes, and this could go on forever! No matter how many primes there are, we have proven there is at least one more!

In this section, we learned about proving by contraction and applied this idea to a few examples.

In the next section, we will learn about proving by induction.

Proof by mathematical induction

Mathematical induction allows us to prove each of an infinite sequence of logical statements, $p_1, p_2, ...$, is true. The argument involves two steps:

- **Basis step**: Prove p_1 is true.
- **Inductive step**: For a fixed $i \geq 2$ value, assume p_{i-1} is true and prove p_i is true.

If both steps are done successfully, the conclusion is that $p_1, p_2, ...$ are all true.

But how can we make this conclusion? The idea is that we have shown p_1 is true and that each p_i is true assuming p_{i-1} is true. Therefore, let $i = 1$, then p_2 is true. Let $i = 2$, then p_3 is true. Let $i = 3$, then p_4 is true. This pattern continues indefinitely, so each p_n must be true.

Mathematical induction can be thought of as an infinite line of dominoes standing on their edges. If you knock one over, it falls into the next, which falls into the next, which falls into the next, and on and on.

This discussion is admittedly a bit abstract, so let's actually do some proofs by mathematical induction to understand how it can be used.

Example – Adding 1 + 2 + ... + n

Suppose we wish to show that, for any positive integer n, the following formula is true:

$$1 + 2 + 3 + \cdots + n = \frac{n(n+1)}{2}$$

We need to show that this is true for all positive integers, $n = 1, 2, 3, \ldots$, so we get an infinite chain of statements we want to prove:

$$p_1 : \quad 1 = \frac{1(1+1)}{2}$$

$$p_2 : \quad 1 + 2 = \frac{2(2+1)}{2}$$

$$p_3 : \quad 1 + 2 + 3 = \frac{3(3+1)}{2}$$

$$\vdots \qquad \vdots$$

$$p_{i-1} : \quad 1 + 2 + 3 + \cdots + (i-1) = \frac{(i-1)((i-1)+1)}{2}$$

$$\vdots \qquad \vdots$$

Figure 2.11 – Adding example

First, the **basis step**. Let $n = 1$; then, the left side of the equation is 1 and the right side is the following:

$$\frac{1(1+1)}{2} = \frac{2}{2} = 1$$

So, the formula is true for $n = 1$. In other words, p_1 is true.

Second, the **inductive step**. Let $n = i - 1$ and assume p_{i-1} is true, which means the following:

$$1 + 2 + 3 + \cdots + (i - 1) = \frac{(i - 1)i}{2}$$

Let's add i to both sides and try to prove p_i is true:

$$1 + 2 + 3 + \cdots + (i - 1) + i = \frac{(i - 1)i}{2} + i$$

$$1 + 2 + 3 + \cdots + i = \frac{(i - 1)i}{2} + \frac{2i}{2}$$

$$1 + 2 + 3 + \cdots + i = \frac{i^2 - i + 2i}{2}$$

$$1 + 2 + 3 + \cdots + i = \frac{i^2 + i}{2}$$

$$1 + 2 + 3 + \cdots + i = \frac{i(i + 1)}{2}$$

At last, the final line is precisely p_i. Therefore, we have proven p_i is true, so, by induction, we have proven the following:

$$1 + 2 + 3 + \cdots + n = \frac{n(n + 1)}{2}$$

This is for any natural number n by the method of mathematical induction.

In summary, we proved p_1 is true for the basis step; that is, the formula is correct for $n = 1$. Then, we assumed the formula is correct for the sum of the first $i - 1$ positive integers for some $i \geq 2$. In other words, we assumed p_{i-1} is true. Next, we showed this assumption implies p_i is true, meaning the formula is correct for the sum of the first i positive integers.

This is the principle of mathematical induction in action, and we are done with the proof because, if we let $i = 2$, the proof from the basis step for $p_{i-1} = p_1$ implies $p_i = p_2$ is true by the inductive step, which implies p_3 is true, which implies p_4 is true, and on and on, so the formula is true for any n.

It is important to realize these proof methods can be used with many types of mathematical structures, not just numbers, so let's consider a more interesting geometric use of the principle of mathematical induction.

Example – Space-Filling Shapes

Suppose we have a grid of squares that is *2n* in length and *2n* in height for some positive integer *n*. Then, we will call the following specific type of octagon a T-gon, although it may be rotated:

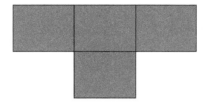

Figure 2.12 – A T-gon is a shape like the letter T

We will seek to prove that any 2^n-by-2^n grid with $n \geq 2$ can be filled with non-overlapping T-gons that will only cover space inside the grid. In other words, T-gons can *tile* the grid. We can break this claim down into a sequence of statements:

- p_2: A 2_2-by-2_2 grid can be tiled with T-gons.
- p_3: A 2_3-by-2_3 grid can be tiled with T-gons.
- p_4: A 2_4-by-2_4 grid can be tiled with T-gons.

And so on as the exponents grow. Note that we are starting at p_2. We could call it p_1 as we have before, but it is easier to simply start at p_2, so the subscript corresponds to *n*.

For the **basis step**, let *n* = 2 so that the grid has the shape $2^2 \times 2^2 = 4 \times 4$. This is a small enough grid that we can easily show that four T-gons can fill this grid—that is, we can prove p_2—as we see here:

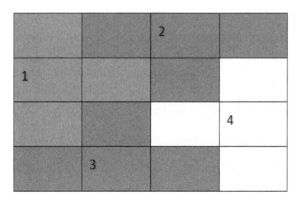

Figure 2.13 – A 4-by-4 grid can be filled with four T-gons rotated as shown here

For the inductive step, let $i \geq 3$ and assume that p_{i-1} is true; that is, a 2^{i-1}-by-2^{i-1} grid can be tiled by T-gons. An important insight is that a *2i-by-2i* grid can be made up of four 2^{i-1}-by-2^{i-1}-adjacent grids aligned in the way displayed in the following diagram:

Grid of dimensions $2^{i-1} \times 2^{i-1}$	Grid of dimensions $2^{i-1} \times 2^{i-1}$
Grid of dimensions $2^{i-1} \times 2^{i-1}$	Grid of dimensions $2^{i-1} \times 2^{i-1}$

Figure 2.14 – One 2i-by-2i grid is made up of four adjacent 2^{i-1}-by-2^{i-1} grids

Now, since p_{i-1} tells us each of these 2^{i-1}-by-2^{i-1} grids can be tiled by T-gons, we can simply tile all four of those in the preceding figure so that the larger *2i-by-2i* grid is tiled by T-gons; that is, p_i is true. In other words, p_{i-1} being true implies p_i is true, so T-gons can tile any grid of dimensions *2n-by-2n* by the principle of mathematical induction, so this completes the proof.

So, we have seen some nice toy examples that are good for understanding the method of proof by mathematical induction, but let's try another problem that has some practical implications for comparing the speeds of algorithms, a topic we will study deeply later in the book.

Example – exponential versus factorial growth

It turns out that different algorithms react differently to having a different number of inputs, usually corresponding to bigger problems. An exponential algorithm with *n* inputs might require the computer to do 2^n arithmetic operations, while a factorial algorithm with *n* inputs might require a number of operations equal to the following:

$$n! = 1 \cdot 2 \cdot 3 \cdots n$$

As *n* grows, the sequences *n!* and 2^n accelerate and grow quickly, but which one grows faster? Does one grow faster for early *n* but slower for larger *n*? This is really not obvious at all. Let's look at a plot of the two sequences to see how they seem to compare:

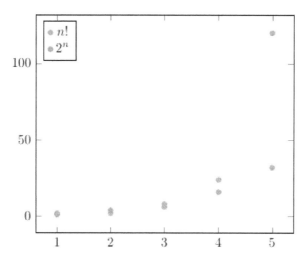

Figure 2.15 – Plots of n! and 2^n

From this plot, we see the factorial sequence surpasses the exponential sequence when $n = 4$, and seems to remain higher, but our plot only goes up to $n = 5$, so it is not obvious what may happen as n continues to grow, so let's try to prove $n! > 2^n$ for $n \geq 4$.

To set up an inductive proof, say we have the following:

$$p_4 : 4! > 2^4$$

$$p_5 : 5! > 2\wedge 5$$

And so on for p^6, p^7, ... Here, we do not look at p^1, p^2, and p^3 because we see $n!$ is smaller than 2^n for these values. We are more interested in when the comparison with n gets larger. Keep in mind that we still have an infinite sequence of p_i statements to prove, but we simply start at p_4.

For the basis step, we can easily see that $4! = (4)(3)(2)(1) = 2^4$ and $2^4 = 16$, so we have $4! > 2^4$, which confirms p_4 is true.

For the inductive step, we will assume p_{i-1} is true for some value $i \geq 5$, which means the following:

$$(i - 1)! > 2^{i-1}$$

Let's try to prove p_i is true. Multiply each side of the inequality by i to get the following:

$$i(i-1)! > 2^{i-1} \cdot i$$
$$i! > 2^{i-1} \cdot 5$$
$$i! > 2^{i-1} \cdot 2$$
$$i! > 2^i$$

Thus, p_i is true. Therefore, $n! > 2^n$ for all $n \geq 4$ by mathematical induction.

A conclusion we can make is that a factorial time algorithm is slower than an exponential time algorithm for any reasonably large problem because the number of computations required will be higher for factorial time algorithms, in fact much higher as n grows. This is a topic we'll study in much more detail in *Chapter 7*, *Computational Requirements for Algorithms*.

Summary

In this chapter, we introduced the primary results of formal logic and proved logical statements by using truth tables. We also learned about constructing mathematical proofs using several methods, such as direct proofs, proofs by contradiction, and proofs by mathematical induction. In addition, these different methods for constructing mathematical proofs were accompanied by simple step-by-step examples to help you think like a mathematician and use deductive thought, which will be helpful for the rest of the chapters in this book.

In the next chapter, we will learn about numbers in base n and perform some arithmetic operations with them. We will also learn about binary and hexadecimal numbers and their uses in computer science.

3
Computing with Base-n Numbers

We are all accustomed to decimal (base-10) numbers. In this chapter, we will introduce numbers in other bases, describe arithmetic operations with those numbers, and convert numbers from one base to another. We will then move to binary digits (base-2), which are the foundation on which all computer operations are built, develop an approach to efficient arithmetic with them, and look at some of the core uses of binary, including Boolean algebra. Lastly, we will discuss hexadecimal (base-16) numbers and their uses in computer science. We will use Python code to do some computations such as converting decimal numbers to binary and hexadecimal and use Boolean operators to select and view data that satisfies a certain criterion.

In this chapter, we will be covering the following topics:

- Base-n numbers
- Converting between bases
- Binary numbers and their application
- Boolean algebra
- Hexadecimal numbers and their application

By the end of this chapter, you should be able to write numbers in different bases and convert numbers from one base to another. For example, 123 is a base-10 number that can be converted into other bases, depending on the need. You will also learn about the importance of binary and hexadecimal number systems along with their applications in computer science.

> **Important Note**
> Please navigate to the graphic bundle link to refer to the color images for this chapter.

Understanding base-n numbers

In this section, we will discuss how to write numbers in different bases with the help of some examples.

A base-n system uses n different symbols for writing numbers, as in $0, 1, 2, ..., n - 1$. This n is called the *radix* of the numbering system. Of course, the customary base-10, or decimal, numbers use the digits 0 through 9.

All base-n numbers make use of the *positional* system, like the one used by decimal numbers, which we will discuss in the next example.

Example – Decimal numbers

Let's think about what it means to write the decimal number 3214 with the usual positional system. It seems trivial, but it is important to realize what exactly a digit in each position in this number represents in order to understand the commonality between the base-10 system we all know and this new idea of a base-n system. The number is made up of a sum of three thousands (10^3), two hundreds (10^2), one ten (10^1), and four ones (10^0), which we can write as follows:

$$3214 = 3 \cdot 10^3 + 2 \cdot 10^2 + 1 \cdot 10^1 + 4 \cdot 10^0$$

To distinguish between numbers written in different radixes, the radix is written as a subscript after the number. For example, 3214 in base-6 form is written as $(3214)_6$. If no radix is specified, it is assumed to be in decimal (base-10) form unless the context makes some other base clear. As we can see, this number represented by this sequence of digits in base-6 has a very different value than the same sequence of symbols in decimal.

There is an unlimited number of different base-n systems, as we could theoretically use any real number for n, but only certain systems have been used frequently in applications. Some of the most widely used ones are noted in the following table:

Name	Radix (n value)	Symbols used
Binary	2	0 and 1
Trinary	3	0, 1 and 2
Quadrary	4	0, 1, 2 and 3
Octal	8	0, 1, 2, 3, 4, 5, 6 and 7
Decimal	10	0, 1, 2, 3, 4, 5, 6, 7, 8 and 9
Hexidecimal	16	0, 1, 2, 3, 4, 5, 6, 7, 8, 9, A (10), B (11), C (12), D (13), E (14) and F (15)

Figure 3.1 – Common base-n numbering systems

Note that when we have bases larger than 10, we can no longer simply use a subset of the digits 0 through 9. For example, the hexadecimal system, which is commonly used in a number of applications in computer science, needs 16 distinct symbols, so it uses 0 through 9 and also the letters A through F. These letters represent the equivalent of the decimal numbers 11 through 15. We will learn about the hexadecimal number system later in this chapter.

Definition – Base-n numbers

A non-negative integer number can be represented in base-n as follows:

$$(d_k d_{k-1} \cdots d_1 d_0)_n,$$

Here, the digits d_0 through d_k are not multiplied, but just written side by side. The decimal value of this number is this:

$$d_k n^k + d_{k-1} n^{k-1} + \cdots + d_1 n^1 + d_0 n^0$$

Now that we have a definition of base-n numbers and we have seen some examples, we can think about what it means to convert between different bases.

Converting between bases

Now that we have the basic knowledge about base-n numbers, let's move on and see how these numbers transform between different bases. We can transform numbers in any base to base-10 and vice versa. In this section, we will show the conversion between different bases along with examples and Python code.

Converting base-n numbers to decimal numbers

Using the definition of base-n numbers given previously, we can convert the following numbers to base-10, or decimal, form. Several examples follow:

- $(a)_n = a \cdot n^0 = a$
- $(ab)_n = a \cdot n^1 + b \cdot n^0 = an + b$
- $(abc)_n = a \cdot n^2 + b \cdot n^1 + c \cdot n^0 = an2 + bn + c$
- $(abcd)_n = a \cdot n^3 + b \cdot n^2 + c \cdot n^1 + d \cdot n^0 = an^3 + bn^2 + cn + d$

We can apply this according to the number of digits we have.

Example – Decimal value of a base-6 number

Let's convert the number $(3214)_6$ into decimal form for this example:

$$(3214)_6 = 3 \cdot 6^3 + 2 \cdot 6^2 + 1 \cdot 6^1 + 4 \cdot 6^0 = 648 + 72 + 6 + 4 = 730$$

This is far from the decimal number 3214. We can see that the same number (here 3214) has a different value based on the base it is written in. The most-used base is base-10.

Base-n to decimal conversion

To convert a decimal number to a certain base n, we repeatedly divide the number at hand by n and keep track of the remainders as we proceed with the division. Let's illustrate this procedure with the help of an example.

Example – Decimal to base-2 (binary) conversion

Let's convert 357 into binary form.

We repeatedly divide 357 by 2 and keep track of the remainders. First, we divide 357 by 2 to get 178 with a remainder of 1, which we write on the right side of the following table. In the next row, we divide 178 by 2 to get 89 with no remainder (0). We continue this in each row until we are unable to do it anymore:

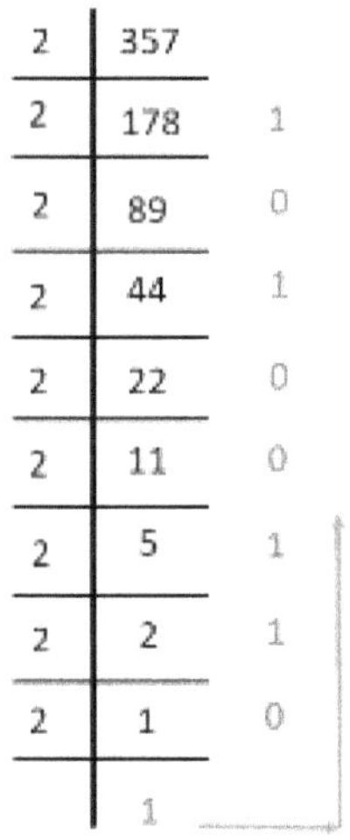

```
2 | 357
2 | 178  1
2 |  89  0
2 |  44  1
2 |  22  0
2 |  11  0
2 |   5  1
2 |   2  1
2 |   1  0
    |   1
```

Figure 3.2 – Converting a decimal number to binary

Now that we have the divisions performed and the remainders noted down, we can follow the direction of the arrows to get our binary number, that is, $(101100101)_2$. This method can be used to convert a decimal number to any non-decimal base (base-2 for this example).

Now that we know how to do the conversion, let's investigate why this method works. For our current example, in order to convert into base-2, we repeatedly divide by 2, and so the remainders can only be 0 (for even numbers) or 1 (for odd numbers). Hence, a base-2 number only uses 0 and 1 for its representation.

The same goes for numbers represented in other bases. For example, to convert a decimal number to base-7, we would repeatedly divide by 7, and so the possible remainders vary from 0 through 6, which are the digits for representing a base-7 number.

Let's do some more examples to make this clearer.

Example – Decimal to binary and hexadecimal conversions in Python

Let's use Python to convert a decimal number to binary and hexadecimal. When you run the code, it will prompt you to enter a number of your choice, which will then be converted into both binary and hexadecimal numbers:

```
# TypeConversion from decimal with base 10
# to hexadecimal form with base 16
# to binary form with the base 2

# Taking input from user - an integer with base 10
number = int(input("Enter a number with base 10\n"))
# Converting the decimal number input by user to Hexadecimal
print("Hexadecimal form of " + str(number) + " is " +
    hex(number).lstrip("0x").rstrip("L"))
# Converting the decimal number input by user to Binary
print("Binary form of " + str(number) + " is " + bin(number).
    lstrip("0b").rstrip("L"))
```

The output, if the user inputs 12345, is as follows:

```
Enter a number with base 10
123456
Hexadecimal form of 123456 is 1e240
Binary form of 123456 is 11110001001000000
```

From the preceding example, we can see that the hexadecimal number system is shorter and therefore easier to work with as compared to the binary number system.

In this section, we learned about different number systems and how to convert numbers from one base to another.

Next, we will continue to discuss a few applications in computer science of binary (base-2) numbers and hexadecimal (base-16) numbers.

Binary numbers and their applications

In this section, we will learn about the binary number system in detail along with its applications and importance in computer science. In particular, we will consider a brief history of binary, provide an explanation as to why they are so foundational to how computers work, and examine the link between binary numbers and Boolean algebra and its use in databases.

The modern binary number system, which is the basis for binary code, was invented by Gottfried Leibniz in 1689, which he described in his article *Explication de l'Arithmétique Binaire* (translated as "explanation of binary arithmetic").

Binary numbers are represented in a base-2 system. The only digits used to represent a binary number are "0" and "1." Each digit is called a *bit*. A binary string of eight bits can represent any of *256 (2^8)* possible values.

A bit string is not the only kind of binary code; other systems that allow only two choices, such as ON/OFF or True/False, can be binary in nature. One such example is Braille, developed by Louis Braille. Braille is widely used by the blind to read and write by touch. This system consists of grids of six dots each, three per column, in which each dot has two states: raised or not raised. Different combinations of these raised or flattened dots represent different letters, numbers, punctuation, and so on. Here are some examples of how alphabets are written in Braille by making use of raised and flattened dots:

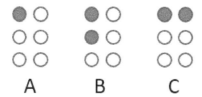

Figure 3.3 – Alphabets in Braille

The importance of the binary number system to the development of computers goes way back to 1946, when the first electric general-purpose digital computer – **Electronic Numerical Integrator and Computer** (**ENIAC**) – was built at the University of Pennsylvania.

The brain of a computer (the CPU) has many circuits that are made up of a large number of transistors. Transistors are analogous to a "switch" that can be turned to the ON or OFF states based on the signal it receives. The binary digits 0 and 1 reflect the OFF and ON states of a transistor. The user provides the computer with a set of instructions for the computer to do a task. These instructions/commands are translated (by a compiler) into binary code for the computer to understand and execute. All the data, information, music, pictures, and so on are processed and stored in binary form by the computer.

As mentioned before, a 0 or a 1 is called a bit. A group of eight bits is called a *byte*. Let's try representing multiples of bytes in the decimal and binary systems:

Multiple of bytes			
Value	Metric system	Value	Binary system
1000	kilobyte (kB)	1024	kibibyte (KiB)
1000^2	megabyte (MB)	1024^2	mebibyte (MiB)
1000^3	gigabyte (GB)	1024^3	gibibyte (GiB)
1000^4	terabyte (TB)	1024^4	tebibyte (TiB)
1000^5	petabyte (PB)	1024^5	pebibyte (PiB)
1000^6	exabyte (EB)	1024^6	exbibyte (EiB)
1000^7	zettabyte (ZB)	1024^7	zebibyte (ZiB)
1000^8	yottabyte (YB)	1024^8	yobibyte (YiB)

Figure 3.4 – Multiples of bytes and their value in metric and binary systems

The binary interpretation of metric prefixes is used by most operating systems.

Boolean algebra

In this section, we will learn about Boolean algebra in detail, along with its applications, such as logic gates. Boolean operators are very useful in filtering out and viewing data that meets certain criterion; this will be illustrated by using Python to solve an example.

George Boole introduced the idea of Boolean algebra in his book titled *The Mathematical Analysis of Logic* in 1847. Boolean algebra is a subset of algebra in which values of variables are either "True" (1) or "False" (0). The main operations of Boolean algebra are detailed here.

The AND operator

This operator states that for an output to occur, two or more events must happen simultaneously. However, the order in which the individual events occur is irrelevant. We use & to represent the AND operator. Hence, we can say that A & B = B & A, which means it agrees with **commutative law**.

Boolean algebra has applications in electronics. Let's try to understand the AND operator by making use of a simple electric circuit comprising a lamp, a battery, and two switches (A and B), as shown in the following figure:

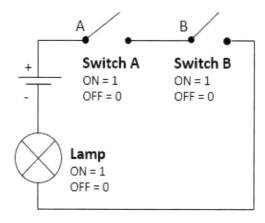

Figure 3.5 – A circuit showing an AND operator

For the lamp to glow, both switches A and B must be in the "ON" (1) position. If either of the switches is ON with the other in the OFF position, then the circuit is incomplete, and the lamp does not glow. The following figure shows the application of Boolean algebra of 0 and 1 to electronic hardware comprising logic gates connected to form a circuit diagram:

Figure 3.6 – AND gates

If A and B are switches, then both must be closed (=1) for the circuit to be closed and the current to flow. If either of the switches is open, then the circuit is open and the current does not flow.

Mathematically, it can be written as $A \wedge B$. If $A=1$ and $B=1$, then $A \wedge B = 1$, otherwise $A \wedge B = 0$. This can be represented by the following figure:

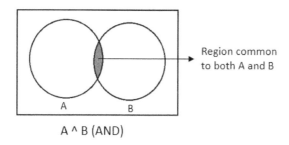

Figure 3.7 – The AND operator

Let's learn about the OR operator in the next section.

The OR operator

This operator states that for an output to occur, either of two conditions needs to be true. Let's try to understand this by making use of electric circuits. In *Figure 3.8*, we can see that the circuit will be closed, and the lamp will glow if either switch A is ON or switch B is ON, or both are ON:

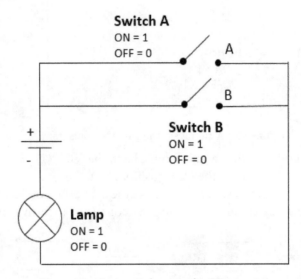

Figure 3.8 – A circuit showing the OR operator

Mathematically, this can be written as *A V B*:

If *A=1, B=1*, then *A V B =1*.

If *A=0, B=0*, then *A V B =0*.

If *A=1, B=0*, then *A V B =1*.

If *A=0, B=1*, then *A V B =1*.

This can be represented by the following figure:

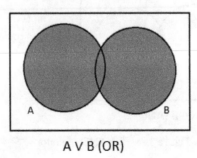

A V B (OR)

Figure 3.9 – The OR operator

The AND and OR operators can be summarized by making use of truth tables as shown in *Figure 3.10*. Here, 0 = False/OFF and 1 = True/ON:

A	B	A ∧ B	A ∨ B
0	0	0	0
0	1	0	1
1	0	0	1
1	1	1	1

Figure 3.10 – A truth table for the AND and OR operators

Now that we know how the OR operator works, we will learn about the NOT operator in the next section.

The NOT operator

This operator is used to reverse the truth value of an entire expression, from False to True or from True to False, depending on the situation.

Let's say that a university wants to send a warning email to students whose GPA is less than 2.0. This statement can be reframed in another way – send a warning email to students whose GPA is *not* greater than 2.0.

This operator can be represented by ¬A:

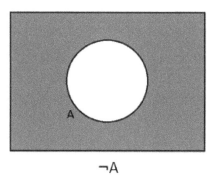

¬A

Figure 3.11 – The NOT operator

The NOT operator is represented in a circuit diagram/logic gate as shown in the following figure:

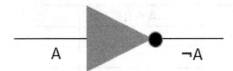

Figure 3.12 – The NOT operator

The NOT operator can be summarized by making use of a truth table as shown in *Figure 3.12*. Here, 0 = False/OFF and 1 = True/ON:

A	¬A
0	1
1	0

Figure 3.13 – A truth table for the NOT operator

Let's see how we can use all this theory about Boolean operators in an example.

Example – Netflix users

Boolean operators can be used to select and view data that satisfies a certain criterion. Let's use the following table to show how our operators can be used in Python. *Figure 3.13* shows the customer addresses for 10 customers of Netflix:

CustomerID	Country	State	City	Zip Code
1	USA	Georgia	Atlanta	30332
2	USA	Georgia	Atlanta	30331
3	USA	Florida	Melbourne	30912
4	USA	Florida	Tampa	30123
5	India	Karnataka	Bangalore	560001
6	India	Maharashtra	Mumbai	578234
7	India	Karnataka	Hubli	569823
8	India	Maharashtra	Mumbai	578234
9	Germany	Bavaria	Munich	80331
10	Canada	Ontario	Toronto	M4B 1B3

Figure 3.14 – Netflix customer dataset

For this example, we will be using a Python library called pandas. It is a fast, flexible, and easy-to-use open source data analysis and manipulation tool that is built on the top of the Python programming language.

> **Important note**
> Installing Python packages, such as pandas in this instance, is an important skill that everyone needs. Here's a link with detailed instructions regarding how to install different packages in Python: `https://packaging.python.org/tutorials/installing-packages/`.

In addition, we will need to import our data to Python in order to use the code in this example. The data is stored in a **Comma-Separated Value** (**CSV**) file provided on GitHub called `CustomerList.csv`, which is available in the GitHub repository for this book. Be sure to download it and store it in the same directory where you store your code.

We will do the following for this example:

- Use the AND operator to view the customers who live in the USA (AND) in the state of Georgia.
- Use the OR operator to view the customers who live either live in the USA or in the state of Ontario.
- Use the NOT operator to view the customers who do not live in the USA.

The Python code is as follows:

```
# Import packages with the functions we need
import pandas as pd

# Import the file you are trying to work with
customer_df = pd.read_csv("CustomerList.csv")

# Using AND operator
print("Example for AND operator")
print(customer_df.loc[(customer_df['Country'] == 'USA') &
    (customer_df['State'] == 'Georgia')])

# Using OR operator
print("Example for OR operator")
print(customer_df.loc[(customer_df['Country'] == 'USA') |
    (customer_df['State'] == 'Ontario')])
```

```
# Using NOT operator
print("Example for NOT operator")
print(customer_df.loc[(customer_df['Country'] != 'USA')])
```

The output is as follows:

```
Example for AND operator
   CustomerID Country    State     City Zip Code
0           1     USA  Georgia  Atlanta    30332
1           2     USA  Georgia  Atlanta    30331
Example for OR operator
   CustomerID Country    State       City Zip Code
0           1     USA  Georgia    Atlanta    30332
1           2     USA  Georgia    Atlanta    30331
2           3     USA  Florida  Melbourne    30912
3           4     USA  Florida      Tampa    30123
9          10  Canada  Ontario    Toronto  M4B 1B3
Example for NOT operator
   CustomerID  Country        State       City Zip Code
4           5    India    Karnataka  Bangalore   560001
5           6    India  Maharashtra     Mumbai   578234
6           7    India    Karnataka      Hubli   569823
7           8    India  Maharashtra     Mumbai   578234
8           9  Germany      Bavaria     Munich    80331
9          10   Canada      Ontario    Toronto  M4B 1B3
```

In the preceding example, we were able to display records that match a certain criterion – the first task was to view the customers that reside in the USA and in the state of Georgia. Records matching both these requirements were then displayed. Similarly, for the second part of the example, we were able to view the records of customers who either live in the USA or in the state of Ontario (in Canada). We used the OR operator to achieve this goal. Lastly, we used the NOT operator to view all the records for customers that do not reside in the USA; all the results except for the ones who reside in the USA were displayed.

In this section, we learned about different kinds of logical operators and how they can be used to search and view results that match a certain criterion. In the next section, we will be discussing another kind of number system, called the hexadecimal number system, and learning about its application.

Hexadecimal numbers and their application

In this section, we will learn about the hexadecimal number system and its application. We use hexadecimal numbers in our day-to-day lives without realizing, such as for the MAC address of your phone or computer.

Hexadecimal numbers are base-16 numbers. They can be represented by using 10 digits (0 to 9) and 6 letters (*A = 10, B = 11, C = 12, D = 13, E = 14, F = 15*).

Let's look at some conversions between the decimal and hexadecimal number systems:

Hexadecimal	Decimal	Hexadecimal	Decimal
0	0	11 = (1 x 16) + 1	17
1	1	12 = (1 x 16) + 2	18
2	2	13 = (1 x 16) + 3	19
3	3	14 = (1 x 16) + 4	20
4	4	15 = (1 x 16) + 5	21
5	5	16 = (1 x 16) + 6	22
6	6	17 = (1 x 16) + 7	23
7	7	18 = (1 x 16) + 8	24
8	8	19 = (1 x 16) + 9	25
9	9	1A = (1 x 16) + 10	26
A	10	1B = (1 x 16) + 11	27
B	11	1C = (1 x 16) + 12	28
C	12	1D = (1 x 16) + 13	29
D	13	1E = (1 x 16) + 14	30
E	14	1F = (1 x 16) + 15	31
F	15	20 = (2 x 16) + 0	32
10 = (1 x 16) + 0	16		

Figure 3.15 – Counting in hexadecimal

Just like decimal numbers, hexadecimal numbers also have place values:

$$(100)^{16} = (1 \cdot 16^2) + (0 \cdot 16^1) + (0 \cdot 16^0) = 256$$

Computer programmers use hexadecimal numbers to simplify the binary number system. We know that $2^4 = 16$, so we know there is a linear relationship between 2 and 16, which implies that four binary digits would be equivalent to one hexadecimal digit. In other words, since binary numbers can be represented by two digits (0 or 1) and hexadecimal numbers can be represented by 16 digits and letters, and we can write 16 as a power of 2 (2^4), four binary digits would be equivalent to one hexadecimal digit. While computers use the binary numbering system, humans use the hexadecimal system to make things easier to understand.

Example – Defining locations in computer memory

In the previous section, we learned that 1 byte = 8 bits. Hexadecimal numbers can characterize every byte as two hexadecimal digits as compared to eight digits when the binary number system is used.

Let's work through an example to better understand how memory locations are defined on a computer, how different variables are stored in different memory locations, and how the values assigned to variables (and, hence, the memory locations) can be changed. We will do the following for this example:

- We will define a `peanut_butter` variable and assign the value 6 to it. We will then print the memory location of where this variable is stored.
- We will define another variable, `sandwich`, and assign it the same value as `peanut_butter`. When we print the memory location of this variable, we will see that it is the same as for `peanut_butter`. This is because we assigned the same value (6) to both our variables, and so they were stored in the same memory location.
- We will move on to assign 7 to the `sandwich` variable and then set both the `peanut_butter` and `sandwich` variables to each other. We can check that they both return the same memory location.
- We'll then set `sandwich` to 10; this changes the value (and memory location) of the `sandwich` variable only, and nothing changes for the `peanut_butter` variable.

Let's see how to implement this in Python:

```
#Variable 1: peanut_butter
peanut_butter = 6
print("The memory location of variable peanut_butter is:
   ",id(peanut_butter))

#Variable 2: sandwich
```

```python
sandwich = 6
print("The memory location of variable sandwich is:
  ",id(sandwich))

print(" We can see that the memory location of both the
  variables is the same because they were assigned the same
    value")

#Setting value of sandwich variable to a new number
sandwich = 7

#Setting both the variables equal to each other:
peanut_butter = sandwich
print("After setting the values of both variables equal to each
  other, we have: ")

print("The value of variable sandwich is now set to:
  ",sandwich)
print("The value of variable peanut_butter is now set to:
  ",peanut_butter)

print("The value of  sandwich variable was changed to 10, let's
  see whether it affects the value of peanut_butter")
#Setting value of sandwich variable to a new number
sandwich = 10

print("The value of variable peanut_butter: " ,peanut_butter)
print("The value of peanut_butter did NOT change even though we
  changed the value of sandwich")
print("The memory location of variable peanut_butter is:
  ",id(peanut_butter))
```

The output of the code is as follows:

```
The memory location of variable peanut_butter is:  2077386960
The memory location of variable sandwich is:  2077386960
 We can see that the memory location of both the variables is
   the same because they were assigned the same value
After setting the values of both variables equal to each other,
we have:
The value of variable sandwich is now set to:  7
The value of variable peanut_butter is now set to:  7
```

```
The value of variable was changed to 10, let's see whether it
  affects the value of peanut_butter
The value of variable peanut_butter:   7
The value of peanut_butter did NOT change even though we
  changed the value of sandwich
The memory location of variable peanut_butter is:   2077386976
```

Now that we know how hexadecimal numbers are used to define memory locations, let's move on to see some more examples of how they are useful.

Example – Displaying error messages

Hexadecimal numbers represent the memory location of errors, making it easier for the user to find and fix them. A binary representation, which would be the most natural representation due to the way a CPU works, would include four times as many digits, which would be difficult for a human to read and interpret.

Example – Media Access Control (MAC) addresses

MAC addresses are unique identifiers assigned to the **Network Interface Card** (**NIC**) of any computer. An NIC is required in order to connect to other computers in a network. It is useful for uniquely identifying a computer among other computers. The format of a MAC address is either AA:AA:AA:BB:BB:BB or AAAA-AABB-BBBB:

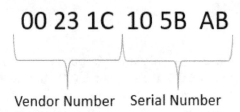

Figure 3.16 – A MAC address

We can easily write some Python code to find the MAC address of the device on which it is running by writing the following code in the terminal:

```
import uuid

# address using uuid and getnode() function
# making use of hexadecimal number system
print (hex(uuid.getnode()))
```

It has the following output:

```
0xf40669da5f06
```

Now that we know how to find the MAC address of our computers, let's move on to see how the hexadecimal number system can be used to define colors.

Example – Defining colors on the web

The primary colors – red (R), green (G), and blue (B) – are represented by two hexadecimal digits each. This can be written as #RRGGBB. Primary colors cannot be created by mixing other colors.

The values of red, green, and blue can be set between 0 and 255 to generate other colors. *Figure 3.17* lists all the commonly used colors:

Color	Red Value	Green Value	Blue Value	Hexadecimal
Red	255 (FF)	0 (00)	0 (00)	#FF0000
Green	0 (00)	255 (FF)	0 (00)	#00FF00
Blue	0 (00)	0 (00)	255 (FF)	#0000FF
Yellow	255 (FF)	255 (FF)	0 (00)	#FFFF00
Orange	255 (FF)	165 (A5)	0 (00)	#FFA500
Aqua	0 (00)	255 (FF)	255 (FF)	#00FFFF
Navy Blue	0 (00)	0 (00)	128 (80)	#000080
Black	0 (00)	0 (00)	0 (00)	#000000
White	255 (FF)	255 (FF)	255 (FF)	#FFFFFF

Figure 3.17 – For each RGB value, we have written the value in decimal and the two-digit hex number in parentheses in columns 2-4

The advantages of hexadecimal number system:

- It's a concise number system, so we can store more information by using less memory space.
- It is more human-friendly because it allows the grouping of binary numbers.

In this section, we learned about hexadecimal numbers and some of their applications, which included defining locations in computer memory, MAC addresses for devices, displaying error messages, and defining colors on a web page.

Summary

In this chapter, we learned about numbers in different bases (decimal, binary, hexadecimal) and how we can convert between bases. Binary numbers are a base-2 number system, whereas decimal numbers are base-10 and hexadecimal numbers are base-16, respectively. We also learned about one very crucial application of the binary number system – Boolean algebra and Boolean operators.

In the next chapter, we will be learning about combinatorics, which includes the study of permutations and combinations that will enable you to calculate the amount of memory required to store certain kinds of data. In addition, we will learn about hashing and the efficacy of brute force algorithms.

4
Combinatorics Using SciPy

This chapter is about counting (or combinatorics), which seems simple, but rapidly gains complexity when counting the number of ways to combine, order, or select various finite sets. This includes the study of permutations and combinations, which can be applied to determining the memory required to store various types of data.

We will apply these ideas to measure the efficacy of brute-force algorithms applied to cryptography and the traveling salesman problem.

In this chapter, we will cover the following topics:

- The fundamental counting rule
- Counting permutations and combinations of objects
- Applications to memory allocation
- Efficacy of brute-force algorithms

By the end of the chapter, you will be able to count various mathematical structures, distinguish between combinations and permutations, and be able to count them. You will also be able to apply these ideas to practical problems in memory allocation and measure the effectiveness of brute-force algorithms in code-breaking in cryptology, the traveling salesman problem, and beyond. The SciPy Python library as well as the standard Python `math` library will be used in this chapter.

> **Important Note**
> Please navigate to the graphic bundle link to refer to the color images for this chapter.

The fundamental counting rule

This section is devoted to counting the number of possible ways to select several objects, each from a set of distinct elements. We will first focus on the case of just two sets before extending it to an arbitrary number of sets.

Definition – the Cartesian product

The set of ordered pairs $A \times B = \{(a, b) : a \in A, b \in B\}$, with component a as an element from set A and the second component b from set B, is called the Cartesian product of sets A and B:

	a_1	a_2
b_1	(a_1, b_1)	(a_2, b_1)
b_2	(a_1, b_2)	(a_2, b_2)

Figure 4.1 – If $A = \{a_1, a_2\}$ and $B = \{b_1, b_2\}$, then $A \times B$ consists of the ordered pairs in this table

This chapter is all about counting the number of elements in sets. Recall from *Chapter 1, Key Concepts, Notation, Set Theory, Relations, and Functions* that the cardinality of a set is the number of elements in the set. Cartesian products are interesting things to count because we can count the number of ways of choosing one element from set A and another element from set B, so our first counting rule will find the cardinality of a Cartesian product.

Theorem – the cardinality of Cartesian products of finite sets

If A and B are finite sets, then $|A \times B| = |A| \cdot |B|$.

Proof: Assume $|A| = n$. If B is the empty set, then $|A \times B| = 0$. If $A = \{a_1, ..., a_n\}$ and $B = \{b_1\}$, then the elements of $A \times B$ are clearly $(a_1, b_1), ..., (a_n, b_1)$ so that $|A \times B| = |A| \cdot |B| = n \cdot 1 = n$.

Suppose $B = \{b_1, ..., b_m\}$, then we can break $A \times B$ down into the disjoint $A \times (B - \{b_1\})$ and $A \times \{b_1\}$ sets, so we have the following from the previous step:

$$|A \times B| = |A \times (B - \{b_1\})| + |A \times \{b_1\}| = |A \times (B - \{b_1\})| + n$$

Repeat this process, deleting one element from B until B runs out of elements (m times) and we eventually have the following:

$$|A \times B| = |A \times \emptyset| + \underbrace{|A| + |A| + \cdots + |A|}_{n \text{ times}} = 0 + nm = nm = |A| \cdot |B|$$

Practically, this result says that if we must choose a pair of elements consisting of one item from a set of m distinct elements and one from another set of n distinct elements, there are mn unique ways to do it. This idea easily extends to situations where there are more than just two sets. First, we extend the definition of the Cartesian product to more sets, which leads to the fundamental counting rule, the key to most of the remainder of our upcoming counting rules.

Definition – the Cartesian product (for n sets)

The set $A_1 \times A_2 \times \ldots \times A_n = \{(a_1, a_2, \ldots, a_n) : a_1 \in A_1, a_2 \in A_2, \ldots, a_n \in A_n\}$ of ordered n tuples, where the ith component, ai, comes from set A_i for each instance of $i = 1, 2, \ldots, n$.

Theorem – the fundamental counting rule

If $A_1, A_2, \ldots, A_n$ are finite sets, then $|A_1 \times A_2 \times \ldots \times A_n| = |A_1| \cdot |A_2| \ldots |A_n|$.

Proof: Assume $|A_i| = m_i < \infty$ for each instance of $i = 1, \ldots, n$. Apply the previous result for calculating a Cartesian product to $A_1 \times A_2$ to find $|A_1 \times A_2| = |A_1| \cdot |A_2| = m_1 m_2$. Then, choose one ordered pair from $A_1 \times A_2$ and pair it with one element from A_3. Then, the same result implies the following:

$$|(A_1 \times A_2) \times A_3| = |A_1 \times A_2| \cdot |A_3| = |A_1| \cdot |A_2| \cdot |A_3| = m_1 m_2 m_3$$

This is the number of elements in $A_1 \times A_2 \times A_3$. Continuing this process, we can include one more set, A_i, repeatedly until we get to A_n and we will have the result of the theorem.

In other words, if we need to select n elements, each from sets of $m_1, m_2, \ldots, m_n$ distinct elements, there are $m_1 m_2 \ldots m_n$ unique ways to do it.

Example – bytes

A byte is a unit of digital information typically consisting of eight bits, which are binary digits: ones and zeroes. We can use the fundamental counting rule to calculate the number of possible bytes that could be constructed. Each of the eight digits may be filled with an element of the set $\{0,1\}$, which has a cardinality of 2; so, we have the following:

$$|\{\text{all possible bytes}\}| = 2 \cdot 2 \cdot 2 \cdot 2 \cdot 2 \cdot 2 \cdot 2 \cdot 2 = 2^8 = 256$$

Therefore, if we suppose each possible byte represents some particular information, each one can carry one of 256 possible pieces of information.

Another option to determine the figure in the preceding example would be to list all the possible bytes and count them: *00000000, 00000001, 00000010, 00000011,* ...; but as you can already tell, this would take quite some time! The takeaway from this remark is that while there are typically "brute-force" approaches to count complex things, using combinatorial rules, including the preceding one, is far more practical.

Example – colors on computers

In many computing applications, colors are created by mixing the colors **red, green, and blue** (**RGB**). In particular, you can specify the intensity of each color in a mixture. A common approach used in HTML and CSS, among other technologies, is to encode the intensity of each color as 1 byte of information.

Let's count how many unique colors this approach can create. Since we have 3 bytes, and each byte can take one of 256 forms, we can see that there are *256 · 256 · 256 = 16,777,216* unique combinations of intensities of red, green, and blue, and so this approach can create over 16 million colors!

As we have seen, the fundamental counting rule directly allows us to compute some quantities of interest, such as the information that can be communicated with bytes and the number of colors that certain web languages can display. Beyond that, it is a key result that will lead to formulas for computing other sorts of groupings of objects: permutations and combinations.

Counting permutations and combinations of objects

This section is dedicated to counting orderings, or permutations, of objects in a set, as well as subsets of specified cardinalities, or combinations, of elements of some wider set.

Definition – permutation

A permutation is a rearrangement of the elements of a set.

Example – permutations of a simple set

For the set *{1, 2, 3}*, the set of all permutations is *{123, 132, 213, 231, 312, 321}*, so there are six permutations of this set. Certainly, there is nothing special about elements 1, 2, and 3. Any set of three distinct elements would have the same number of permutations.

Theorem – permutations of a set

The number of permutations of a set of size n is $n! = n(n-1)(n-2)\ldots(2)(1)$, which is pronounced *n factorial*.

Proof: In the first position of the permutation set, there may be any of the n objects. If we have selected one, that leaves $n-1$ remaining objects for the second position, and so on. According to the fundamental counting rule, there are $n(n-1)(n-2)\ldots(2)(1) = n!$ possible permutations.

Notice that in the previous example, this theorem tells us the number of permutations of $A = \{1, 2, 3\}$ is $|A|! = 3! = 3\cdot 2\cdot 1 = 6$, the same result we found from listing all the permutations.

> **Important Note**
> Note that $0!$ is defined to be 1.

Example – playlists

Suppose we have a playlist of 20 songs that we will play in a random order (without repeating). According to the previous theorem, the number of possible orders, or permutations, is $20! \approx 2.43 \times 10^{18}$, a shockingly high number!

Growth of factorials

Notice in the following table that factorials grow extremely quickly:

n	$n!$	n	$n!$
1	1	6	720
2	2	7	5,040
3	6	8	40,320
4	24	9	362,880
5	120	10	3,628,800

Figure 4.2 – A table of the first 10 factorials. As we see, a set of 10 elements has over 3 million permutations!

The number of permutations of just 10 elements is over 3 million. By the time we reach 20 elements, as in the previous example, the number of permutations is $20! \approx 2.43 \times 10^{18}$, over 2 quintillion!

Many computational tools, such as calculators and programming languages, cannot (by default) calculate permutations if the number of elements gets too high, but the factorial function from the math module in Python (`math`) does not experience much trouble, as it can calculate large factorials efficiently using some mathematical tricks, as seen in the following code:

```
import math
print(math.factorial(20))
print(math.factorial(100))
```

The resulting output is as follows:

```
2432902008176640000

93326215443944152681699238856266700490715968264381621468592963895217599993229915608941463976156518286253697920827223758251185210916864000000000000000000000000
```

> **Important Note**
> The `math` module will be used frequently in this book. Check out the official documentation for the `math` module at https://docs.python.org/3/library/math.html for more details.

Sometimes, we may wish to count a slightly different type of permutation; for example, in our example with playlists, suppose we want to randomly play only half the playlist of 20 songs. Then, how many distinct permutations of a subset of 10 of the 20 songs are there? The next result allows us how to easily calculate that number.

Theorem – k-permutations of a set

The number of permutations of k out of n distinct elements from a set, or k-permutations, is as follows:

$$P_k = \frac{n!}{(n-k)!}$$

Proof: If there are k positions to fill from n options, the first position may be filled by any of the n elements, the second may be filled by any of the remaining $n - 1$ elements, and so on down to the last position being filled by any of the remaining $n - k + 1$ elements. So, according to the fundamental counting rule, we have $n(n - 1)(n - 2)\cdots(n - k + 1)$ possibilities, but this can be manipulated as follows:

$$n(n-1)(n-2)\ldots(n-k+1) = n(n-1)(n-2)\ldots(n-k+1) \cdot \frac{(n-1)(n-k-1)\ldots(2)(2)}{(n-k)(n-k-1)\ldots(2)(1)}$$

This gives a fraction with $n!$ as the numerator and $(n - k)!$ as the denominator, or $\frac{n!}{(n-k)!}$

This theorem works for all non-negative integers, $k \leq n$, so we see that the previous theorem is just a special case of this more general theorem where $k = n$.

Returning to our playlist, then, the number of 10-permutations of the 20-song list can be computed with Python as follows:

```
import math
print(math.factorial(20)/math.factorial(20-10))
```

This outputs the following:

```
670442572800.0
```

Now, we can count the number of permutations or orderings of sets or their subsets. Another sort of grouping of elements is a combination, which we will discuss next.

Definition – combination

A combination is a selection of some elements from a set.

The main difference between a combination and a permutation of some k out of n elements is that different orderings of the same k elements represent multiple permutations, but only one combination.

Example – combinations versus permutation for a simple set

Consider the set $A = \{1, 2, 3\}$. The two-element permutations of A are $\{12, 21, 13, 31, 23, 32\}$, but the set of two-element combinations of A are $\{12, 13, 23\}$. Since the order is not important for combinations, we have fewer of them. The next result shows just how many we have.

Theorem – combinations of a set

The number of combinations of k out of n elements of a set, or k-combinations, is

$$C_k = \binom{n}{k} = \frac{n!}{k!\,(n-k)!}.$$

Proof: The number of permutations of k out of n elements is $_nP_k$ by the theorem on k-permutations. For each fixed k elements, there are $k!$ different permutations by the first theorem on permutations, so we simply need to divide $_nP_k$ by $k!$ to find the number of combinations of k out of n elements, since the order of the elements does not matter in combinations. So, we have the following:

$$\frac{P_k}{k!} = \frac{n!}{k!\,(n-k)!} = \binom{n}{k}$$

Binomial coefficients

$\binom{n}{k}$ is called a binomial coefficient because of the binomial theorem from algebra, which gives the expansion of a binomial raised to the power of a non-negative integer n, as follows:

$$(x+y)^n = \sum_{k=0}^{n} \binom{n}{k} x^k y^{n-k} = \binom{n}{0} x^0 y^n + \binom{n}{0} x^1 y^{n-1} + \cdots + \binom{n}{n} x^n y^0$$

Example – teambuilding

Suppose there are 20 software engineers working in an office. Their supervisor will choose a team of four engineers to work on a new project. We would like to count the number of possible teams that could be selected. Note that the order in which the team members are selected is unimportant to counting the number of teams—for example, the team of Katie, Pranav, Sanjay, and Li is the same as the team of Pranav, Li, Sanjay, and Katie. Therefore, the correct structures we are counting are combinations rather than permutations.

Therefore, the number of possible teams is $\binom{20}{4}$, which would be cumbersome to calculate by hand, so we can use a tool such as Python. We could use the factorial function from before, along with the definition of binomial coefficients, but there is a highly optimized implementation in the SciPy package, specifically in its special functions, called `binom`:

```
# using the factorial function
import math
print(math.factorial(20) / math.factorial(4) / math.
```

```
    factorial(20-4))

# import the special functions from sciPy
import scipy.special
print(scipy.special.binom(20,4))
```

The output is shown here:

```
4845
4845.0
```

Therefore, there are 4,845 distinct teams that could be chosen for the project. Note that both of the code examples work, but `scipy.special.binom` is preferable because it is optimized.

> **Important Note**
> You can find the official documentation for the popular SciPy library for Python at https://docs.scipy.org/doc/scipy/reference/.

Example – combinations of balls

Consider a box containing six red balls and five yellow balls and assume five balls are to be chosen randomly. Let's find the number of combinations where there are exactly three red balls in the five chosen.

First, we have to choose three out of the six red balls, so there are $\binom{6}{3} = 20$ ways of doing that. Secondly, we must choose two of the five yellow balls, so there are $\binom{5}{2} = 10$ ways of doing that. We need to choose one of the 20 ways of getting the correct number of red balls and one of the 10 ways of selecting the correct number of yellow balls, so according to the fundamental counting rule, there are $20 \cdot 10 = 200$ ways that both of these can occur.

As we see, to solve more complicated problems, several of the combinatorial rules we have established may be needed.

In the remainder of the chapter, we will discuss some practical applications of combinatorics in computer and data science.

Applications to memory allocation

One area where combinatorics can come into play is in determining how much memory an algorithm needs to complete a certain task. It is frequently useful to know this before we run some code. In most programming languages, when arrays are created, they are given a static size that cannot be changed. Therefore, it is faster or more convenient to change an existing value in an array than to change the size of an array.

So, developers often pre-allocate the memory by creating an array of the maximum size we will need for the whole course of the algorithm, either filled with 0s or empty, depending on the language. This is not a problem with small amounts of data, but when the program needs to process exponentially large amounts of data, this can be very wasteful. Understanding memory usage is also important to avoid certain negative consequences: we may use up so many resources on the device that it cannot complete its other tasks, it may crash, or it may begin reading and writing data to hard drives instead of the much faster RAM.

Example – pre-allocating memory

Suppose we wish to create a large list of *1,000,000* numbers. The simplest way is to just run a loop, adding one element at a time to the vector:

```
import time
number = 1000000

# Check the current time
startTime = time.time()

# Create an empty list
list = []

# Add items to the list one by one
for counter in range(number):
    list.append(counter)

# Display the run time
print(time.time() - startTime)
```

It returns the following:

```
0.584686279296875
```

Therefore, this code runs in about 0.5847 seconds, which seems fast, but is not optimal.

> **Important Note**
>
> The Python time library allows you to measure the runtime of some code. The time.time() command checks the current time, so if you save this value at the beginning, you can measure the time elapsed by subtracting that from a new time.time() command at the end.
>
> The runtime will depend on the computing device, so you may find a different amount of time than the preceding example.

Suppose we pre-allocate a list of length 1,000,000 with the following code before filling it in:

```
import time
number = 1000000

# Check the current time
startTime = time.time()

# Create a list of 1000000 zeros
list = [None]*number

# Add items to the list one by one
for counter in range(number):
    list[counter] = counter

# Display the run time
print(time.time() - startTime)
```

We get an output as follows:

```
0.44769930839538574
```

The runtime here is only *0.4477* seconds, a time saving of 23%. Here, we readily see the speed advantage of pre-allocation, at least for large lists in Python. Sure, saving 0.14 seconds is inconsequential on a small scale, but if you use this method in an algorithm that will run thousands or millions of times in some software, it can make a huge difference.

While each requires a loop of 1,000,000 iterations, Python must do more work in the first method as each iteration requires more operations to be done—according to the fundamental counting rule, each extra operation turns into 1 million more operations upon completing all million iterations—so it takes more time, even though the two approaches both accomplish the same goal. Furthermore, most languages are even less efficient than Python at repeated appending elements to lists.

It should be mentioned that while a list of size 1,000,000 may seem large, this not at all an uncommon size for the objects we may analyze. For example, a typical photo taken with a modern smartphone may include more than 10 million pixels, each of which would have three numbers associated with it (RGB value), so a list of 30 million numbers would be needed to represent a single picture file. As you might imagine, a video file may include thousands of pictures, leading to enormous list sizes.

In the next section, we will learn about brute-force algorithms and go through some examples, such as the Caesar cipher and the traveling salesman problem.

Efficacy of brute-force algorithms

A combination lock requires you to input three numbers from, say, *0* to *9* to open the lock. One approach to open it if you forget the password is to try *(0, 0, 0)*, then *(0, 0, 1)*, then *(0, 0, 2)*, and so on. This method is guaranteed to succeed if we have enough patience to test all permutations of *0* through *9* for each of the three numbers. This is a brute-force algorithm: a trial-and-error approach to solving a problem where you simply guess the answer over and over until you get it right. Of course, this is very tedious for a combination lock, but brute-force approaches are actually sometimes practical, especially when using computers.

Example – Caesar cipher

Roman emperor and general Julius Caesar is said to have been one of the earliest users of encryption in the form of coded messages. Now called the Caesar cipher, his method was to write the message and then shift the alphabet by some specified number of letters. For example, he might choose to shift the alphabet by *4* letters. Then, *A* is replaced by *E*, *B* is replaced by *F*, and so on. When we reach *V*, it becomes *Z*. After that, we go back to the beginning so that *W* becomes an *A*, *X* becomes a *B*, and so on, as we see here:

plaintext	A	B	C	...	U	V	W	X	Y	Z
ciphertext	E	F	G	...	Y	Z	A	B	C	D

Figure 4.3 – The plaintext characters and corresponding ciphertext characters

One approach to breaking some encryption is brute force. If we know it was a Caesar cipher, then there are only *25* possible shifts, so we can just test them all until we find one that seems right. We will do this in the following code:

```
# Intercepted message
codedMessage = 'nzohfu gur rarzl ng avtug'

# We will shift by 0, shift by 1, shift by 2, ... and print the
  # results
for counter in range(26):
    # Start with no guess
    guessedMessage = ''

    # Loop through each letter in the coded message
    for x in codedMessage:

        # If x is not a space
        if x != ' ':

            # Shift the letter forward by counter
            if ord(x)+counter <= 122:
                x = chr(ord(x)+counter)

            # Subtract 26 if we go beyond z
            else:
                x = chr(ord(x)+counter-26)

        # Build a guess for the message one letter at a time
        guessedMessage = guessedMessage + x

    # Print the counter (the shift) and the message
    print(counter, guessedMessage)
```

A few lines of the output are as follows:

```
10 xjyrpe qeb bkbjv xq kfdeq
11 ykzsqf rfc clckw yr lgefr
12 zlatrg sgd dmdlx zs mhfgs
13 ambush the enemy at night
14 bncvti uif fofnz bu ojhiu
```

We see that the cipher must have shifted the alphabet by 13, as we discover by inspecting each of the possible adjusted alphabets.

Moreover, this example shows something about when brute-force algorithms are useful. For brute-force algorithms to thrive, there are two main requirements:

- The set of possible solutions, or solution space, is sufficiently small.
- It must be possible to determine the correct solution given output from each possible solution.

If we fail condition 1, the algorithm takes too long to run. In fact, the problem with such large solution spaces is that it would take days or even years to run a brute-force algorithm. In this example, however, there were only 26 possible answers. It is important to note that for each answer, we had to execute several operations, but the overall runtime is quite small on a modern computer. If we fail condition 2, we will not know whether we have found the right answer even if we have it. It's obvious here because most of the strings of text are not intelligible messages, so we can pick out Caesar's message right away.

In this section, we will focus on the first conditions because we can use combinatorics to count the sample space for various problems to evaluate the efficacy of brute-force algorithms.

Staying on the theme of cryptanalysis (the art and science of breaking codes), suppose we receive the following encrypted message:

```
toa bxfew grknm cks jxuyz kdar h lhvp akq
```

If we input this into the brute-force algorithm (try it!), you will see that none of the 26 shifts makes an intelligible message, so the author has apparently used a different, possibly more sophisticated method to encrypt their message. None of the 26 alternative alphabets allowed by the Caesar cipher accurately model the encryption used. A wider class of encryption is a so-called simple substitution cipher wherein each letter in the true message (the plaintext) is replaced by another letter in the coded message (the ciphertext) but is not necessarily a Caesar cipher where the alphabet is just shifted. This leads to a problem that is more like a cryptogram that you might see in a newspaper or puzzle book. A valid brute-force algorithm would have to search a larger set of alternate alphabets and we would have to view the messages to determine whether they are intelligible. But how large is this set of alphabets? Clearly, we can note the following:

- *A* could be replaced by any of the 26 letters of the alphabet.
- *B* could be replaced by any of the 25 remaining letters.
- *C* could be replaced by any of the 24 remaining letters.

A familiar trend emerges: the factorial. Indeed, any of these possible ciphertext alphabets are re-orderings or permutations of the normal alphabet, so there are *26! ≈ 4.03 . 10²⁶*, or *403* heptillion such alphabets.

Clearly, a brute-force algorithm that constructs the messages found by applying each possible alphabet in the solution space and inspecting them manually is not practical. If we could check *10* messages per second, it would take *1.2 quintillion* years, or *10 million* times the age of the universe! A fully computerized version at best may read and check with dictionaries to see whether the text forms words at a rate of several million per second, but this still requires a runtime of billions of years.

Although the brute-force approach of testing would almost certainly produce the right answer if completed, this is not enough for it to be practical. It has to be possible to complete it in a useful period of time. Even worse for brute force, this is not even a complex type of encryption—making up an alphabet by hand can be done in just a couple of minutes!

Example – the traveling salesman problem

Suppose a traveling salesman will drive around to visit N cities, including his home city, to try to sell his wares and then return home. He wants to minimize the distance he travels so that his fuel costs are as small as possible; so, the question of the **Traveling Salesman Problem** (**TSP**) is as follows:

Given the list of cities and the minimum distance between each two cities, in what order should the salesman visit each city and return home with a minimum travel distance?

The TSP is a classical problem in operations research, and we will study more advanced approaches to the problem in *Chapter 9*, *Searching Data Structures and Finding Shortest Paths*; but for now, let's see how this problem responds to a brute-force algorithm by thinking about what exactly must be done to solve it, and also consider the size of the data structures that should be stored.

First, if we have a list of N cities and the distances between two cities, how many distances will there be? To find the number of unique pairs of cities, we would need to consider every combination of 2 out of N cities. Note that we do not consider permutations because, for example, the distance from Chicago to Dallas is the same as the distance from Dallas to Chicago, so the order does not matter, and storing separate distances would be redundant. Thus, the number of distances we will have is as follows:

$$\binom{N}{2} = \frac{N!}{(N-2)!\,2!} = \frac{N(N-1)}{2}$$

This allows us to know precisely how much memory the data will take, allowing pre-allocation of a data structure with space for *(N(N – 1))/2* spaces.

Next, a brute-force way to solve the problem is to simply find the distance of each possible circuit the salesman could make through the cities and compare the distances; so how many such circuits are there? If the salesman starts in his hometown, he has *N – 1* cities to choose from for the second city. After visiting the second city, he has *N – 2* choices for his third city, and so on, until he runs out of cities, when he returns home. This is a permutation, so there are *(N – 1)!* possible circuits he could take.

There is a bit of a problem with this accounting. Suppose there are only five cities and he takes a circuit that we label as follows:

$$a \to d \to e \to c \to f \to b \to a$$

We have highlighted this in red in the following figure, showing the full set of links between the cities:

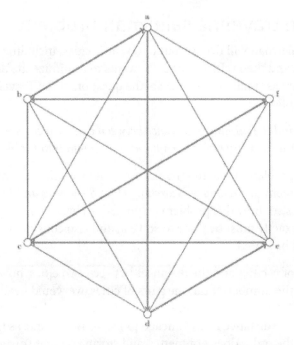

Figure 4.4 – The whole set of paths with the circuit we mentioned in red for a small TSP with N = 6

Another possible circuit is as follows:

$$a \to b \to f \to c \to e \to d \to a$$

This is the same circuit but in the reverse order (simply reverse the arrows in the preceding figure). This circuit will require him to travel the exact same distance as the first circuit since it traverses all the same roads, just in the reverse order. Since the reverse of each circuit will be included in our calculation of *(N – 1)!* circuits, we can divide that by 2 to cut the work that the brute-force algorithm must do in half, as we will need to test only the following:

$$\frac{(N-1)!}{2}$$

This is a substantial reduction in the time needed to carry out such a brute-force approach.

For a six-city problem, there are only *(6)(5)/2 = 15* distances and only *5!/2 = 60* possible circuits the salesman could take (ignoring reverse versions of circuits).

However, even though this reduces the time by 50%, we saw with the previous cryptology problems that brute-force algorithms are only feasible for tiny problems, so reducing the computation by half is not enough to practically solve the TSP unless the number of cities, *N*, is very small. If *N = 20*, still a relatively small problem, we have the following number of possible non-redundant circuits:

$$\frac{19!}{2} \approx 6.08 \times 10^{16}$$

This would be entirely infeasible to solve with brute force.

Summary

In this chapter, we primarily discussed how to count the cardinality, or size, of sets of different types. First, we looked at counting Cartesian products, where we take one element from each of a sequence of sets to create a new set. Counting the size of these comes down to the fundamental counting rule, which we used to count binary structures and the colors that can be displayed with HTML/CSS.

Second, we looked at permutations and combinations using factorials (for permutations) and binomial coefficients (for combinations), which we derived directly from the fundamental counting rule. For factorials, the key tool in Python is the `factorial` function in the `math` library and, for binomial coefficients, the `binom` function from the SciPy library.

Lastly, we took a look at just a few applications of combinatorics in computer science, including memory allocation, the (poor) speed of brute-force algorithms in a few examples in the area of cryptology, and for a classical optimization problem, the TSP.

The tools from this chapter will be used repeatedly as we progress through the book. In particular, counting is important in computing probabilities in the next chapter, *Chapter 5, Elements of Discrete Probability*. The so-called complexity analysis of algorithms will be covered more generally and more deeply in *Chapter 7, Computational Requirements for Algorithms*, and we will continue from the discussion of brute-force algorithms on to more effective approaches for various problems.

5
Elements of Discrete Probability

Probability is the study of randomness, chance, and uncertainty. We experience randomness all the time–from the weather to the stock market to the results of sporting events and elections. We can never predict these things with certainty, but we can make reliable statements about the likelihood (or probability) of events occurring through careful study of patterns in the uncertainty and variables that may affect it.

The type of probability that's most important to discrete mathematics and computer science is to do with, of course, discrete sets. In this chapter, after establishing how probability works in the general sense, we will present elements of combinatorial probability. This is important in situations where each resulting outcome of a random experiment is equally likely, so that the chance that the result is in a certain set of outcomes which depends on counting the size of the set. Then, we will look at conditional probability and Bayes' theorem, which allow us to update probabilities based on learning new information, which is quite an important idea in machine learning and other topics. We will then use this theory to consider Bayesian spam filters, which try to automatically identify which emails are legitimate and which are not. The key turns out to be Bayes' theorem, which takes in user input when it makes mistakes in classifying emails and updates its approach to improve over time.

After that, we will discuss random variables, which take some random numerical values and analyze them by considering their average values through the idea of a mean of a random variable and how erratic they are via the idea of their variance. All of this will culminate in a look at Google's PageRank system for ranking search results, which revolutionized web searches in the late 1990s and early 2000s.

In this chapter, we will be covering the following topics:

- The basics of discrete probability
- Conditional probability and Bayes' theorem
- Bayesian spam filtering
- Random variables, means, and variance
- Google PageRank I

> **Important Note**
> Please navigate to the graphic bundle link to refer to the color images for this chapter.

The basics of discrete probability

As we have said, making predictions or finding probabilities requires careful analysis, so we need a mathematical framework for probability. It will all center around the idea of a random experiment.

Definition – random experiment

A random experiment is any process that has an uncertain outcome.

Simple examples of random experiments are tossing a coin or rolling a die, each of which has an uncertain outcome. These are easy to analyze, but some random experiments are much more difficult, such as predicting tomorrow's weather. Despite the complexity, experts can estimate the chance of each possible result of the random experiment using complex meteorological models, taking into account temperatures, humidity, and other atmospheric data.

Something each example has in common is that there is a random result for each experiment. A coin toss may result in heads or tails. We may roll a 1, 2, 3, 4, 5, or 6 on the die. The weather may be clear tomorrow, or it may rain or snow. These are called outcomes.

Definitions – outcomes, events, and sample spaces

Let's look at what outcomes, events, and sample spaces are:

- Each possible result of a random experiment is an outcome.
- A set of outcomes is an event.
- A sample space S is the set of all possible outcomes of a random experiment.

Example – tossing coins

Consider a random experiment where we toss a coin. Let H represent the coin landing on heads and let T represent the coin landing on tails. The sample space of this random experiment is $S = \{H, T\}$.

The coin can land on heads or tails, each of which is a single outcome. Events are sets of outcomes, that is, subsets of S. All possible events would be $\emptyset$, $\{H\}$, $\{T\}$, and $\{H, T\}$.

Example – tossing multiple coins

Instead of just one coin, consider a random experiment where we toss three coins. In this case, the outcome of the experiment is a sequence of three outcomes from several coin tosses. Therefore, the sample space S consists of the following:

TTH	TTH	THT	THH
HTT	HTH	HHT	HHH

Figure 5.1

The list of all possible events for this random experiment would be quite long. Keep in mind that events are simply any subsets of the 8 outcomes in the sample space shown previously. This makes for a total number of events given here:

$$\binom{8}{0} + \binom{8}{1} + \binom{8}{2} + \binom{8}{3} + \binom{8}{4} + \binom{8}{5} + \binom{8}{6} + \binom{8}{7} + \binom{8}{8} = 256$$

We know this from *Chapter 4*, *Combinatorics Using SciPy*, which was on combinatorics for counting combinations.

As you may have noticed, randomly flipping a coin is equivalent to randomly selecting a binary digit—often *0* represents tails and *1* represents heads, as we will see in some more advanced examples later—which feeds nicely into computer science applications due to the ubiquity of binary.

With our coverage of the ideas of random experiments and their sample spaces, we have established all the things that could occur from some random process, but not the core quantity we seek: the chance of each outcome occurring. As you might suspect, each random experiment has its own way of assigning these values to events; a function takes events as inputs and returns probabilities. Such a function is called a probability measure.

Definition – probability measure

A probability measure is a function $P: \{Events\} \to [0,1]$ mapping events to numbers between 0 and 1 (probabilities), where $P(S) = 1$ and the countable additivity holds.

For pairwise-disjoint events $A_1, A_2, \ldots$, we have $P(A_1 \cup A_2 \cup \ldots) = P(A_1) + P(A_2) + \ldots$.

> **Important Note**
> This means, for every pair of events A_n and A_m from the sequence, $A_n \cap A_m = \emptyset$ if $m \neq n$. In other words, the events are non-overlapping events; they share no outcomes in common.

Let's unpack this definition a little.

The codomain of any probability measure P is $[0,1]$. The outputs are probabilities, or chances of events occurring, so they should not be more than 100% or less than 0%. The higher this output, the more likely the event is to occur:

> **Important Note**
> If an event has a probability of 0, it is not true in general that an event cannot occur, but this is true in the context of discrete probability for finite sets. Likewise, a probability of 1 does not imply an event must occur in general.

1. The probability of the whole sample space $P(S)$ is 1. The sample space consists of all the possible outcomes, so the probability that one of them occurs must be 1.
2. The countable additivity property says that if some events (sets of outcomes) are disjointed (do not share any outcomes), then we can calculate the probability that one event of the group occurs as the sum of their individual probabilities.

From these definitions, we can easily arrive at some elementary properties of probabilities.

Theorem – elementary properties of probability

Let A and B be events, then $P(A \cup B) = P(A) + P(B)$ if A and B are disjointed:

1. $P(\emptyset) = 0$.
2. $P(A^c) = 1 - P(A)$

Proof

The preceding theorem can be proven as follows.

Since A and B are disjointed, $A \cup B$ is just a simpler version of the set $A_1 \cup A_2 \cup \ldots$ from the countable additivity condition of the definition of a probability measure, so the same result applies—namely, $P(A \cup B) = P(A) + P(B)$:

1. Since $S = S \cup \emptyset$ and these sets are disjointed, the previous result and the fact that $P(S) = 1$ gives us $P(S) = P(S) + P(\emptyset)$, or $1 = 1 + P(\emptyset)$, so $P(\emptyset) = 0$.
2. Notice S is a union of the two disjoint sets, A and A^c; by the previous result, $P(S) = P(A \cup A^c) = P(A) + P(B)$. Then, we have $1 = P(A) + P(A^c)$, or $1 - P(A) = P(A^c)$.

All of these properties are intuitive results:

- The first property says the probability that event A or event B happens is the sum of the probabilities when they share no outcomes.
- The second property says the probability that there is no outcome is 0—by definition, the random experiment has some outcome, although it is uncertain.
- The third property says the probability that event A does *not* occur is *1* minus the probability that it does occur. As an obvious example of the third property, if there is a 40% chance that it will rain tomorrow, there must be a 60% chance that it will not rain.

Example – sports

The soccer teams Real Madrid CF and FC Barcelona will be competing in an upcoming match. A sports analyst has forecast that Madrid has a 40% chance of winning, Barcelona has a 50% chance of winning, and that otherwise a draw will occur. So, then, what is the probability that a draw will occur?

The first step in many probability problems is to introduce some notation. Let $B = \{FC\ Barcelona\ wins\}$, $M = \{Real\ Madrid\ CF\ wins\}$, and $D = \{a\ draw\ occurs\}$, whose union makes up the sample space S.

What is the probability that a draw will occur? Notice that $D = (B \cup M)^c$; that is, a draw is the complement of Barcelona or Madrid winning the match. So, we have this:

$$P(T) = P(B \cup M)^c = 1 - P(B \cup M)$$

by property 3 above. Next, B and M are disjoint events since both teams cannot win, so property 1 implies this:

$$P(B \cup M) = P(B) + P(M) = 0.5 + 0.4 = 0.9,$$

This means the following:

$$P(T) = 1 - 0.9 = 0.1,$$

So, there is a 10% chance that there will be a draw, assuming the predictions of the analyst are accurate.

Of course, this example is rather simple and could be solved more informally quite quickly using simple intuition, but constructing some suitable notation and referring back to the specific properties of probabilities becomes more and more essential as the complexity of our problems increases.

The next two theorems are fundamental properties of probability that are necessary for some of the more complex results we will consider later.

Theorem – Monotonicity

If $A \subseteq B$, then $P(A) \leq P(B)$.

Proof

Notice that $B = A \cup (B - A)$ (the blue portion plus the orange portion in the figure), which are clearly disjoint sets:

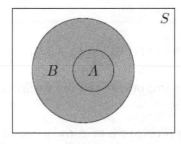

Figure 5.2

Then, the previous theorem tells us that $P(B) = P(A) + P(B - A) \geq P(A)$, since $P(B - A)$ is a probability and, therefore, cannot be negative.

In other words, the property of monotonicity simply means that if we start with some discrete event A and it is possible to add some outcomes to it to create another discrete event B, the probability of event B is the same (if all the extra outcomes have zero probability) or will increase (if the outcomes have positive probability).

Some previous theorems show us how to calculate the probability of a union of *disjoint* events $A \cup B$, but what if A and B share some outcomes? The Principle of Inclusion-Exclusion provides a path to calculating these types of probabilities.

Theorem – Principle of Inclusion-Exclusion

For two events A and B, $P(A \cup B) = P(A) + P(B) - P(A \cap B)$.

Proof

Notice from the diagram that $A \cup B$ consists of three disjoint parts: the orange, blue, and yellow subsets:

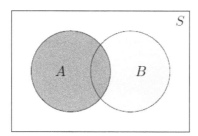

Figure 5.3

Now, $P(A)$ is the sum of the probabilities of the orange and blue parts while $P(B)$ is the sum of the probabilities of the yellow and blue parts. If we were to add these all together, we would add the probability of the blue part, $A \cap B$, twice rather than just once. So, if we subtract one back away, we get $P(A \cup B) = P(A) + P(B) - P(A \cap B)$.

This result gives us a new capability. It allows us to calculate the probability of a union of events, even if the events are not disjoint, with a simple formula.

Definition – Laplacian probability

A Laplacian random experiment is one where every outcome has the same probability.

This verbal description seems rather simple, but when combined with the properties of probability measures, it actually contains much that is instructive.

Theorem – calculating Laplacian probabilities

Consider a Laplacian random experiment:

1. The sample space is finite, $|S| = |\{s_1, s_2, ..., s_n\}| = n < \infty$.

   The probability of each outcome is $\frac{1}{n}$.

   The probability of an event $E \subseteq S$ is $\frac{|E|}{|S|}$.

Proof

We will prove the three claims in order as follows:

1. Let the sample space be a countable (but possibly infinite) set, $S = \{s_1, s_2, ...\}$. Since the experiment is Laplacian, the probability of each outcome, $P(\{s_j\}) = c$ for some number c for every $j = 1, 2,...$, then $P(S) = c + c + ... = \infty$, but this must be 1 if P is a probability measure, which contradicts the assumption that S may be countably infinite, so S must be finite.

2. By the previous result, $S = \{s_1, s_2, ..., s_n\}$ for some finite number n. So, we have this:

$$1 = P(S) = P(\{s_1\}) + \cdots + P(\{s_n\}) = nP(\{s_j\})$$

$$\frac{1}{n} = P(\{s_j\}),$$

   This is equivalent to $1/|S|$.

3. Let $E = \{s_{i_1}, ..., s_{i_k}\} \subseteq S$ where $k \leq n$. Then, we have the following:

$$P(E) = P(\{s_{i_1}, \cdots, s_{i_k}\}) = P(\{s_{i_1}\}) + \cdots + P(\{s_{i_k}\}) = kP(\{s_j\}) = \frac{k}{n} = \frac{|E|}{|S|}.$$

Example – tossing multiple coins

From a previous example, the sample space for tossing three coins is the following:

TTH	TTH	THT	THH
HTT	HTH	HHT	HHH

Figure 5.4

Now, clearly, each of these is equally likely to occur (assuming the coin is fair), so it is a Laplacian random experiment. Then, we see that the probability of each sequence of coin results is 1/8. With this fact, we can calculate some other probabilities:

$$P(\{0 \text{ heads}\}) = P(\{TTT\}) = \frac{1}{8}$$

$$P(\{1 \text{ heads}\}) = P(\{TTH, THT, HTT\}) = \frac{3}{8}$$

$$P(\{2 \text{ heads}\}) = P(\{THH, HTH, HHT\}) = \frac{3}{8}$$

$$P(\{3 \text{ heads}\}) = P(\{HHH\}) = \frac{1}{8}$$

The previous example was pretty simple because we could easily list the whole sample space and count the sizes of the events, but calculating probabilities for Laplacian events with much larger sample spaces requires us to use the combinatorial properties we learned about in the previous chapter.

Definition – independent events

Events A and B are independent if $P(A \cap B) = P(A)P(B)$.

Practically speaking, events A and B do not affect one another. For example, tossing heads on one coin is independent of tossing tails on the next coin.

Example – tossing many coins

Suppose we toss 50 fair coins. By the fundamental counting rule, the sample size here would be $|S| = 2^{50} = 1,125,899,906,842,624$, since each sequence of heads and tails of the 50 coins has 50 parts, each with two possible results.

Of course, the sample size is too large to list it quickly, but we can still calculate probabilities. Suppose we want to know the probability that we get *25 heads*—again, writing down all the events where this occurs is impractical, but we can view the set of sequences where there are exactly *25 heads* as the number of combinations of *50 elements* where *25* are heads; so, we have this:

$$E_{25} = |\{25 \text{ heads}\}| = \binom{50}{25} = 126{,}410{,}606{,}437{,}752$$

That implies the following:

$$P(E_{25}) = \frac{|E_{25}|}{|S|} = \frac{126{,}410{,}606{,}437{,}752}{1{,}125{,}899{,}906{,}842{,}624} \approx 0.1123.$$

Calculating one of these by hand is easy, but calculating the probabilities of E_1, E_2, ..., E_{50} is pretty slow by hand, so let's use Python to compute the binomial coefficients for each index *1, 2, ..., 50* in a loop via SciPy's `binom` function and print out the probabilities of each possible number of heads:

```
# Import packages with the functions we need
import scipy.special
import matplotlib.pyplot as plt

probabilities = []

for n in range(51):
    # Calculate probability of n heads
    probability = scipy.special.binom(50, n) / (2 ** 50)

    # Convert to a string with 6 decimal places
    probString = "{:.6f}".format(probability)

    # Print probability
    print('Probability of ' + str(n) + ' heads: ' + probString)

    # Add probability to list
    probabilities.append(probability)

# Plot the probabilites
plt.plot(range(51), probabilities, '-o')
plt.axis([0, 50, 0, 0.15])
plt.show()
```

This is the (truncated) output:

```
Probability of 22 heads: 0.078826
Probability of 23 heads: 0.095962
Probability of 24 heads: 0.107957
Probability of 25 heads: 0.112275
Probability of 26 heads: 0.107957
```

Note that $P(\{25 \text{ heads}\}) \approx 0.1123$, as we just found. The code also generates a plot with the last three lines of code, as we see here:

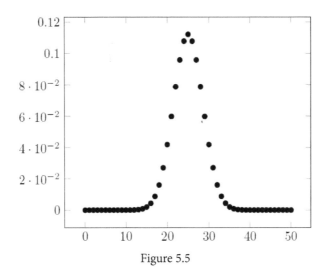

Figure 5.5

Conditional probability and Bayes' theorem

In everyday life, our knowledge of the past informs our predictions about the future. For example, if the team with the best record in a basketball league were about to play against the team with the worst record, we would likely estimate the chance of the first team winning the game to be higher than if we did not know that fact.

This same idea in the context of this chapter would be to calculate the probability of an event occurring after learning that another event has occurred. This is a conditional probability and it applies in situations where we learn information over time, which influences our evaluations of probabilities for subsequent events, which is important to machine learning, artificial intelligence, and many other fields.

Definition – conditional probability

For two events A and B where P(B) > 0, the conditional probability of A given B is as follows:

$$P(A|B) = \frac{P(A \cap B)}{P(B)}$$

This is the proportion of the time A occurs given the knowledge that B also occurs.

Example – temperatures and precipitation

Suppose we have gathered data on high temperatures and whether or not it rained in Melbourne, FL, on May 11 for each year from 1977 to 2018 and have found the following data on high temperatures and the frequency of rain within each temperature category:

Temperatures	Frequency	Rain Frequency
51-60	4	1
61-70	12	5
71-80	13	10
81-90	20	8
91-100	3	1
Totals	50	25

Figure 5.6

We assume the relationship between precipitation and temperature is not significantly changing over time. Suppose a temperature sensor in a particular location is not working, but we are able to detect that it rained—based on this, what is the probability that the temperature is in each range?

Say B = {it rains} and T let be the temperature:

$$P(51 \leq T \leq 60 | B) = \frac{P(\{51 \leq T \leq 60\} \cap B)}{P(B)} = \frac{\frac{1}{50}}{\frac{25}{50}} = \frac{1}{25}$$

Similarly, we end up with this:

$$P(61 \leq T \leq 70|B) = \frac{P(\{61 \leq T \leq 70\} \cap B)}{P(B)} = \frac{\frac{5}{50}}{\frac{25}{50}} = \frac{5}{25}$$

$$P(71 \leq T \leq 80|B) = \frac{P(\{71 \leq T \leq 80\} \cap B)}{P(B)} = \frac{\frac{10}{50}}{\frac{25}{50}} = \frac{10}{25}$$

$$P(81 \leq T \leq 90|B) = \frac{P(\{81 \leq T \leq 90\} \cap B)}{P(B)} = \frac{\frac{8}{50}}{\frac{25}{50}} = \frac{8}{25}$$

$$P(91 \leq T \leq 100|B) = \frac{P(\{91 \leq T \leq 100\} \cap B)}{P(B)} = \frac{\frac{1}{50}}{\frac{25}{50}} = \frac{1}{25}$$

While our faulty temperature sensor makes finding the high temperature impossible, the preceding calculation gives us a probability that the temperature is in each range given the fact that it rained—and a pretty high probability, *0.72*, that the temperature is between *71* and *90*.

Next, we will establish a few more useful results about probability and illustrate how they can be applied with some examples.

Theorem – multiplication rules

If *A* and *B* are events, then the following statements are true:

1. If *P(B) > 0*, then *P(A ∩ B) = P(B)P(A|B)*.
2. If *P(A) > 0*, then *P(A ∩ B) = P(A)P(B|A)*.

Proof

For claim (1), by definition of conditional probability, $\mathbb{P}(A|B) = \frac{\mathbb{P}(A \cap B)}{\mathbb{P}(B)}$. Multiplying both sides by *P(B)* gives *P(B)P(A|B) = P(A ∩ B)*. The result for the second claim follows by the same argument if we interchange the roles of events *A* and *B*.

Note that we previously gave a simpler formula to compute P(A ∩ B) if events A and B are independent, that is, by simply multiplying them, but the formula from this theorem works in any case and, therefore, gives us the capability of calculating some new sorts of probabilities.

Theorem – the Law of Total Probability

Let $A_1, A_2, \ldots$ be events that partition the sample space S. Let B be an event:

$$P(B) = \sum_{n=1}^{\infty} P(A_n \cap B) = \sum_{n=1}^{\infty} P(B|A_n)P(A_n).$$

Proof

Since S is broken into disjoint sets $A_1, A_2, \ldots$ and B is a subset of S, disjoint parts of B are in $A_1, A_2, \ldots$ as well, and $B = (A_1 \cap B) \cup (A_2 \cap B) \cup \ldots$. By countable additivity, P(B) is the sum of their probabilities. The rightmost part of the equation result uses the multiplication rule to rewrite each $P(A_n \cap B)$ as $P(B \mid A_n)P(A_n)$.

The Law of Total Probability is very valuable because it can give us the probabilities of some event B given its probability conditioned on a sequence of other events.

That is, $A_i \cap A_j = \emptyset$ for all $i \neq j$ and $\bigcup_{i=1}^{\infty} A_i = A_1 \cup A_2 \cup \cdots = S$.

Theorem – Bayes' theorem

Let A and B be events with positive probabilities (that is, $P(A) > 0$ and $P(B) > 0$):

$$P(B)P(A|B) = P(A)P(B|A)$$

Equally, the following applies:

$$P(A|B) = \frac{P(B|A)P(A)}{P(B)}$$

Proof

Equating the two results of the previous theorem and dividing each side by *P(B)*, we get this:

$$P(A \cap B) = P(A \cap B)$$

$$P(B)P(A|B) = P(A)P(B|A)$$

$$P(A|B) = \frac{P(B|A)P(A)}{P(B)}.$$

The proof of Bayes' theorem is extremely simple from the definition of conditional probability, but it is nevertheless one of the most important results in all of probability theory, especially in applications where we gain information related to random experiments over time that we want to use to update our evaluation of probabilities, such as continuous video feeds in computer vision, stock prices over time, and more.

Bayesian spam filtering

Suppose we have a filter that flags emails that it identifies as spam. Consider the events *F = {e-mail flagged as spam}* and *T = {e-mail is spam}*. If you have ever used a spam filter, you know that this is imperfect, so these sets do not coincide. Sometimes legitimate messages are caught by a spam filter and sometimes spam is undetected by the filter.

Suppose the developers of the spam filter did some extensive testing on a huge sample of emails and found several results:

- The probability that spam emails will be caught by the filter (true positives) is 0.95, or *P(F|T) = 0.95*.

- The probability that legitimate e-mails are not caught by the filter (true negatives) is 0.98, so *P(F^c|T^c) = 0.98*.

- The probability that an email from the selected sample is spam is 0.1, or *P(T) = 0.1*.

Suppose an email is caught by the filter—what is the probability that it is actually spam? In other words, what is *P(T|F)*? By Bayes' theorem, it would be this:

$$P(T|F) = \frac{P(T)P(F|T)}{P(F)}$$

We do not know the probability that an arbitrary email will be flagged, $P(F)$, but we can use the Law of Total Probability to find it:

$$P(F) = P(F|T)P(T) + P(F|T^C)P(T^C)$$

And, since

$$P(F|T^c) = 1 - P(F|T) = 1 - 0.95 = 0.05$$

and

$$P(T^c) = 1 - P(T) = 1 - 0.1 = 0.9,$$

we have the following:

$$P(T|F) = \frac{(0.1)(0.95)}{(0.95)(0.1) + (0.05)(0.9)} \approx 0.68$$

Therefore, even if an email is flagged as spam, there is only a 68% chance the email is spam given the flaws in the filter, which seemed quite modest at first.

In this section, we have shown how Bayesian probability is commonly used in identifying spam email messages. Spam filtering is one example of a classification problem, which in general try to automatically classify objects into categories. The same general idea is very common in many other classification problems, and Bayesian probability is one of the main tools in this area.

Next, we continue to some more useful probability theory about random variables, which we will combine with the Bayesian ideas we have learned to analyze one of the more influential ideas in the internet era—Google's PageRank algorithm.

Random variables, means, and variance

Informally, we can say that random variables are functions that map outcomes to numerical values. Since the probability measure assigns probabilities to outcomes and events, we may define the probability that a random variable equals certain values. The technical definition is as follows.

Definition – random variable

A function $X: S \to R$, where R is a discrete set, is a discrete **random variable** (**RV**).

> **Important Note**
> The other main class of RVs is continuous RVs, which take values in R or some other uncountable set instead of just a discrete set, but they are outside the scope of this book.

Example – data transfer errors

Data transferred over digital communication channels are, at the lowest level, a stream of binary digits. Sometimes there can be noise or other distortions that cause errors in their transmission. It is important to quantify the errors, but it is random, so the best we can do is estimate the chance of different numbers of errors.

Suppose we send a single byte of eight bits, where each digit has a probability p of being in error and they are all independent of each other. So, what is the probability that some number k out of 8 bits received are incorrect?

By independence, the probability that the first k bits are incorrect and the remaining $8 - k$ bits are correct is $p^k(1-p)^{8-k}$, since the chance of accuracy is $1 - p$. However, the positions of the k errors could be chosen from the 8 bits in $\binom{8}{k}$ ways, so if X is an RV counting the number of errors, then we have the following for $k = 1, 2, \ldots, 8$:

$$P(X = k) = \binom{8}{k} p^k (1-p)^{8-k}$$

Generally speaking, this type of RV is called a binomial RV and the function forms its PMF.

On the other hand, some RVs may not come from some well-known class and may be constructed from empirical data, as the next example shows.

Example – empirical random variable

Consider a 10-sided die with numbers 1 through 10, but it is shaped irregularly with some sides larger than others and an unknown weight distribution, and we would like to know the chance that it takes each value.

Let X be an RV corresponding to the value rolled on the die. To estimate the PMF, one approach is to just roll the die repeatedly and count the number of times it lands on each number. Suppose we roll the die 1,000 times and we get the following frequencies:

Value	1	2	3	4	5	6	7	8	9	10
Frequency	129	242	53	16	57	95	228	33	101	46
Proportion	0.129	0.242	0.053	0.016	0.057	0.095	0.228	0.033	0.101	0.046

Figure 5.7

These proportions serve as an empirical estimate of the PMF of X.

Definition – expectation

Let $X: S \to \{r_1, r_2, \ldots\}$ be a discrete random variable. The expectation of X is defined as follows:

$$E[X] = r_1 f(r_1) + r_2 f(r_2) + \cdots$$
$$= r_1 P(X = r_1) + r_2 P(X = r_2) + \cdots$$
$$= \sum_{i=1}^{\infty} r_i P(X = r_i)$$

If the sum is not infinite, $E[X]$ is also called the expected value or mean of X. Furthermore, if g is a function, then we have the following:

$$E[X] = \sum_{i=1}^{\infty} g(r_i) \mathbb{P}(X = r_i)$$

Note that the expected value is just like a weighted average.

Example – empirical random variable

Continuing with the previous example, we can calculate the expected value of the RV X representing the result of rolling the die as follows:

$$E[X] = (1)(0.129) + (2)(0.242) + (3)(0.053) + (4)(0.016) + (5)(0.057) + (6)(0.095) + (7)(0.228) \\ + (8)(0.033) + (9)(0.101) + (10)(0.046)$$

$$= 4.92$$

So, the die will be valued at 4.92 on average.

The mean of an RV is important because it tells us the average value of the RV if we were to run the underlying random experiment over and over, but this is not the only thing we would typically like to know about an RV.

For example, betting $100,000 on rolling a 1, 2, 3, or 4 on a fair six-sided die would result in the gambler gaining $34,000 on average, but that does not mean it is a good idea! The result is either +$100,000 or -$100,000, and nothing in between, so the RV only takes values far away from the mean.

As this example shows, another important consideration is how much the RV tends to vary from the mean, so we have a measurement of how spread - out the RV is, called variance.

Definition – variance and standard deviation

Let $X: S \to R$ be a discrete random variable; its variance is then this:

$$\sigma^2 = Var(X) = E[(X - E[X])^2]$$

The standard deviation of X is this:

$$\sigma = \sqrt{\sigma^2}$$

The following result is typically the more practical formula to use for calculating variance than the definition given previously.

Theorem – practical calculation of variance

If X is a discrete RV, then the following applies:

$$Var(X) = E[X^2] - E[X]^2$$

Proof

By definition, the following is true:

$$Var(X) = E[(X - E[X])^2]$$

$$= \sum_{i=1}^{\infty} (s_i - E[X])^2 P(X = s_i)$$

$$= \sum_{i=1}^{\infty} s_i^2 P(X = s_i) - 2E[X] \sum_{i=1}^{\infty} s_i P(X = s_i) + E[X]^2 \sum_{i=1}^{\infty} P(X = s_i)$$

$$= E[X^2] - 2E[X]E[X] + E[X]^2$$

$$= E[X^2] - E[X]^2.$$

Example – empirical random variable

Continuing the previous example with an irregular 10-sided die, we can calculate the variance, recalling that $E[X] = 4.92$. First, we calculate this:

$$E[X^2] = (1^2)(0.129) + (2^2)(0.242) + (3^2)(0.053) + (4^2)(0.016) + (5^2)(0.057) + (6^2)(0.095)$$
$$+ (7^2)(0.228) + (8^2)(0.033) + (9^2)(0.101) + (10^2)(0.046)$$

$$= 32.74,$$

That gives us this:

$$Var(X) = E[X^2] - E[X]^2 = 32.74 - (4.92)^2 = 8.5336$$

Google PageRank I

In the late 1990s, there were many search engines on the internet, including Yahoo, Altavista, and Ask Jeeves, but when Google emerged in the early 2000s, it quickly supplanted all of those as the most popular search engine and has remained popular for nearly 20 years, in large part because its results were of such high quality that users flocked to the website. Google used a new approach to web searches that generated very good results.

Developed by Stanford University students, and later Google founders, Larry Page and Sergey Brin (along with researchers Rajeev Motwani and Terry Winograd) in 1996, the algorithm used was called PageRank. Google's primary searching algorithms have certainly progressed from this since 1996 but it remains a key part of their approach.

The key idea of PageRank is to not merely to look for websites that match the user's search terms most closely like most other search tools at the time but to weight the matches by the importance of matching websites in some sense, so that important websites are ranked highest and show up first in the list of search results. They measure importance by counting the number of links and the quality of the links to various web pages. So, the more links a web page has from high-ranked web pages, the higher PageRank will rank the page.

While ingenious, PageRank is actually a fairly simple use of probability. It is easy to understand the main idea of PageRank with the ideas we have developed in this chapter. Suppose we have an internet I of N web pages:

$$I = \{W_1, W_2, ..., W_N\}$$

On I, we define two functions:

1. Outgoing links, $C: I \to \{0, 1, 2, ..., N - 1\}$, where $C(W_j)$ is the number of links leaving the jth web page, where self-links do not count and multiple links to the same web page count as a single link.

2. PageRank, $PR: I \to [0,1]$, where $PR(W_j)$. It is calculated as follows:

$$PR(W_j) = \frac{1-d}{N} + d \sum_{W_i \in M(W_j)} \frac{PR(W_i)}{C(W_i)},$$

Here $M(W_j)$ is the set of web pages linking to W_j. In other words, PageRank is $\frac{1-d}{N}$ plus d times the sum of ratios of PageRank to outgoing links for each other web page linking to W_j.

The constant $d \in (0,1)$ is called the damping factor. (The authors set $d = 0.85$ in their original paper, although Google may have adjusted it since then.) Regardless of the value of d, it can be shown that the function PR is a probability mass function, assigning probabilities to $W_1, W_2, ..., W_N$. (Note that, by definition, the probabilities assigned by a probability mass function sum to 1, so this is what the previous sentence claims, mathematically speaking.)

> **Important Note**
> Note that there is some confusion in the literature about the first term of the *PR* calculation: sometimes *N* is left out of the denominator. This does not have an important impact, but the resulting PageRanks do not form a probability mass function without this *N*.

These probabilities have a more intuitive interpretation. PageRank proposes an imaginary person navigating this internet who randomly click links and will eventually stop on a certain web page. The value d represents the probability that this person will click the next link at each step. The PageRank of a web page $PR(W_i)$ represents the probability that this randomly clicking surfer will stop on web page W_i.

As an example, suppose $N = 5$. In other words, suppose our small internet has only five web pages. Of course, this is unrealistic, but it allows us to paint the picture of how PageRank works on a small scale. We will also assume $d = 0.85$. Furthermore, suppose we have the following structure of links between the web pages:

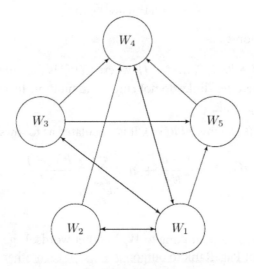

Figure 5.8

- W_1 links to W_2, W_3, W_4, and W_5, so $C(W_1) = 4$.
- W_2 links to W_1 and W_4, so $C(W_2) = 2$.
- W_3 links to W_1, W_4, and W_5, so $C(W_3) = 3$.
- W_4 links to W_1, so $C(W_4) = 1$.
- W_5 links to W_4, so $C(W_5) = 1$.

Note that the formula for PR(W_j) given previously requires knowledge of every other PR(W_i) for $i \neq j$, so it cannot be calculated directly. The typical approach is to initially assume that each PageRank is equal, or $1/N = 1/5$, and then calculate new PageRanks iteratively using knowledge about links.

In the second iteration, the PageRanks are as follows:

$$PR(W_1) = \frac{1-d}{N} + d\left(\frac{PR(W_2)}{C(W_2)} + \frac{PR(W_3)}{C(W_3)} + \frac{PR(W_4)}{C(W_4)}\right) = 0.34$$

$$PR(W_2) = \frac{1-d}{N} + d\left(\frac{PR(W_1)}{C(W_1)}\right) = 0.07$$

$$PR(W_3) = \frac{1-d}{N} + d\left(\frac{PR(W_1)}{C(W_1)}\right) = 0.07$$

$$PR(W_4) = \frac{1-d}{N} + d\left(\frac{PR(W_1)}{C(W_1)} + \frac{PR(W_2)}{C(W_2)} + \frac{PR(W_3)}{C(W_3)} + \frac{PR(W_5)}{C(W_5)}\right) = 0.38$$

$$PR(W_5) = \frac{1-d}{N} + d\left(\frac{PR(W_1)}{C(W_1)} + \frac{PR(W_3)}{C(W_3)}\right) = 0.13$$

Figure 5.9

We see that web pages W_1 and W_4 would be highest ranked, which makes sense as these are the web pages in the diagram with the most incoming links. Furthermore, we see that the sum of all five PageRank values is 1, as we claimed by referring to PR as a probability mass function. In practice, more iterations would be run using the PageRanks we calculated as inputs to the next step along with updated information on links, which may change over time.

These ideas from probability explain how Google's PageRank algorithm works, but this is certainly not the whole story, as we have only considered a small collection of just four web pages. Scaling PageRank up to the entire internet involves the mathematics of linear algebra. We will cover the essentials of linear algebra in *Chapter 6, Computational Algorithms in Linear Algebra*.

Summary

In this chapter, we have primarily discussed the core ideas of probability theory, and in particular discrete probability. These allow us to calculate the probability that an event will occur, or, in other words, the chance that it will occur. We then applied these ideas to some popular modern innovations.

First, we constructed a probability space, made up of a sample space, a set of events, and a probability measure. The definition of these topics led directly to many elementary properties of probabilities and formulas to compute probabilities of events, such as those made up of unions of events and certain intersections of events. This led to an important class of probability spaces: the Laplacian space, where each outcome is equally likely. This reduces many probability calculations to counting problems, which we learned to solve in *Chapter 4, Combinatorics Using SciPy*.

Then, we considered conditional probability, which is essentially the idea that gaining new knowledge should influence our calculation of probabilities. This idea led to some useful results, including Bayes' theorem and the Law of Total Probability. After establishing these results, we continued to apply them to a classification problem—Bayesian spam filtering—which seeks to automatically categorize emails as legitimate or spam.

Lastly, we established a little more probability theory about RVs, their averages via means, and a measure of how random they are: the variance. These ideas, along with Bayesian probability, allowed us to then discuss the Google PageRank approach to ranking results in web searches. In the next chapter, we will learn about computational algorithms that are used in linear algebra.

Part II – Implementing Discrete Mathematics in Data and Computer Science

This part of the book covers applications of discrete mathematics to core concepts of computer science, including linear algebra; the complexity of algorithms in the worst case and on average; storing and extracting features from graphs, trees, and networks, graph searches; and finding shortest paths on networks.

This part comprises the following chapters:

- *Chapter 6, Computational Algorithms in Linear Algebra*
- *Chapter 7, Computational Requirements for Algorithms*
- *Chapter 8, Storage and Feature Extraction of Graphs, Trees, and Networks*
- *Chapter 9, Searching Data Structures and Finding Shortest Paths*

6
Computational Algorithms in Linear Algebra

This chapter covers standard methods and algorithms of linear algebra commonly used in computer science and machine learning problems. Linear algebra centers on systems of equations, a problem where we need to find a set of numbers that solve not just one equation, but many equations simultaneously, using special types of arrays called matrices. Matrices can directly model tree, graph, and network structures that are central to so many computer science applications and the math behind Google's PageRank, among others, all ideas to which we will apply these ideas in later chapters. Systems of equations are key in regression analysis and machine learning.

We will delve into solving these systems of equations from both geometric and computational perspectives before scaling the methods up to solve larger problems with algorithms in Python, because the huge amount of work you would have to do to solve large problems by hand would be impractical.

The mathematical content of the topics is complete, although it may be a refresher for readers, but the computational algorithms and Python functions are likely new.

The following topics will be covered in this chapter:

- Understanding linear systems of equations
- Matrices and matrix representations of linear systems
- Solving small linear systems with Gaussian elimination
- Solving large linear systems with NumPy

The chapter is mostly dedicated strictly to the mathematics of linear algebra and its algorithms, but they will be applied to practical problems in most of the remaining chapters of the book. By the end of the chapter, you will have an understanding of what systems of equations are, and learn how to solve small problems by hand and large problems with some NumPy functions in Python. In addition, you will learn about matrices and how to do arithmetic with them, both by hand and with Python.

> **Important Note**
> Please navigate to the graphic bundle link to refer to the color images for this chapter.

Understanding linear systems of equations

Equations of two variables whose graphs are straight lines, or linear equations, are one of the core parts of any elementary algebra course. They model simple proportional relationships well, but several linear equations taken at once, perhaps involving more than just two variables, allow for the modeling of much more complex situations, as we will see.

In this section, we discuss these familiar equations and then consider the idea of a **system** of multiple linear equations that we wish to solve all at once. We also define linear equations and systems of linear equations that involve more than just two variables and show how to solve them by hand.

Definition – Linear equations in two variables

A linear equation of the variables x_1 and x_2 is any equation that can be written in the form $a_1 x_1 + a_2 x_2 = b$ for some real numbers a_1, a_2, and b. The solutions of the equation are all ordered pairs $(x_1, x_2) \in R_2$ that satisfy the equation.

Definition – The Cartesian coordinate plane

The Cartesian coordinate plane is the familiar concept of a 2D plane on which we can plot points corresponding to an ordered pair of coordinates (x_1, x_2). The first coordinate x_1 represents the horizontal position of the point and the second coordinate x_2 represents the vertical position of the point, as can be seen here:

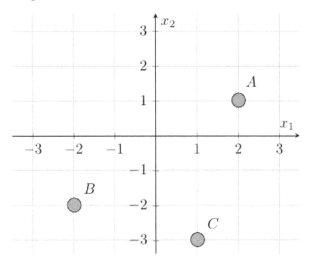

Figure 6.1 – A Cartesian coordinate plane with points A (2,1), B (-2,-2), and C (1,-3)

Some readers may be accustomed to seeing the coordinate axes labeled as x and y with coordinate (x, y), but we go with x_1 and x_2 because we will continue to develop some useful theory in more dimensions.

For example, in 3D space, where we have not only left-right and up-down axes, but also a forward-backward axis, which we will label x_3, additional dimensional spaces are difficult, if not impossible, to visualize fully since our human eyes are adapted to see in the three spatial dimensions, but a 4D space has a fourth axis labeled x_4, a 5D space has a fifth one, and so on.

Example – A linear equation

Consider the linear equation $6x_1 + 3x_2 = 3$. We can solve x_2 in terms of x_1 as follows. First, subtract $6x_1$ from each side to get $3x_2 = 3 - 6x_1$.

Then, divide each side of the equation by 3 to get $x_2 = 1 - 2x_1$.

Therefore, we get the solution set of the equation to be the set of all ordered pairs (in other words, points on the plane), where $x_2 = 1 - 2x_1$, which we can write in set notation as $\{(x_1, 1 - 2x_1) : x1 \in R\}$. In other words, for any given real x-coordinate x_1, we can construct a corresponding y-coordinate as $1 - 2x_1$.

Notice there are infinitely many solutions to the linear equation, one ordered pair for each real number. While we cannot plot infinitely many points to draw the graph of the function in reality, if we choose several x_1 coordinates, compute the corresponding x_2 coordinates, and plot the points on the Cartesian coordinate plane, we see that they are aligned along a linear path:

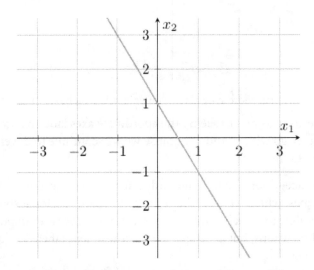

Figure 6.2 – The graph of the linear equation $x_2 = 1 - 2x_1$. Note that the line passes through points $(x_1, x_2) = (1, -1)$ and $(x_1, x_2) = (0, 1)$, which we can see clearly satisfy the equation

It turns out that the graph of every linear equation traces out a straight line in the Cartesian coordinate plane, which is precisely why we call them **linear**.

Definition – System of two linear equations in two variables

A linear system of two equations of variables x_1 and x_2 is made up of two linear equations of x_1 and x_2. A solution to the system is an ordered pair (x_1, x_2) that satisfies both equations simultaneously.

Since each equation can be represented as a line, the geometric equivalent of this problem is to find point(s) of intersection of the lines. Intuitively, it is clear two lines may cross at exactly one point, the lines may be parallel and never intersect, or the lines may coincide with one another entirely.

If the lines cross, we call the system **consistent**. If the lines are parallel, we call the system **inconsistent**. If the lines coincide, we call the system **dependent**. The next three examples will investigate each of these three situations.

Example – A consistent system

Consider the following system of two linear equations:

$$2x_1 + 3x_2 = -1$$
$$6x_1 + 3x_2 = 3$$

To find a solution to the system, we need to find coordinates x_1 and x_2 such that *both* equations are satisfied simultaneously. So, suppose these coordinates exist, then we can think about what must be true about them. The second equation must be true, so if we solve it for x_2 (as we did earlier), we see $x_2 = -2x_1 + 1$, an expression of x_2 in terms of x_1. If we knew x_1, this would provide a formula for us to establish the other coordinate, x_2.

Since the first equation must also be satisfied for a solution (x_1, x_2), it must be valid to replace x_2 with $-2x_1 + 1$ in that equation, which provides a path to find x_1:

$$2x_1 + 3(-2x_1 + 1) = -1.$$

Multiplying the 3 by each term in the parentheses, we have the following:

$$2x_1 - 6x_1 + 3 = -1.$$

Combining the x_1 terms and subtracting 3 from each side of the equation, we have the following:

$$-4x_1 = -4$$

This gives the value of x_1 if we divide each side of the equation by -4:

$$x_1 = 1$$

Thus, if there exists a solution, its x_1 coordinate is 1, but we know we can compute x_2 as

$$x_2 = -2(1) + 1 = -2 + 1 = -1,$$

so, we see the solution must be $(x_1, x_2) = (1, -1)$. Plotting the two lines on a graph confirms that the point $(1, -1)$ is precisely where the two graphs of the linear equations cross:

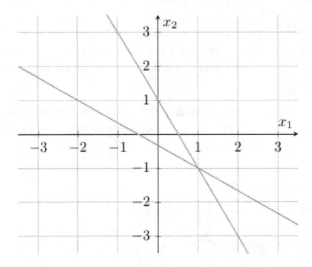

Figure 6.3 – The graphs of the two linear equations cross at point $(1, -1)$

Since the lines cross at one point, it is a consistent system. The point $(1, -1)$ is the only solution to the system of equations, the single point where the lines cross.

Example – An inconsistent system

Consider the following system of two linear equations:

$$2x_1 + x_2 = 3$$
$$6x_1 + 3x_2 = 3$$

To find a solution to the system, again, we need to find coordinates x_1 and x_2 such that *both* equations are satisfied simultaneously. So, suppose these coordinates exist, then we can think about what must be true about them. The second equation must be true, so if we solve it for x_2 (as we did above), we see

$$x_2 = -2x_1 + 1,$$

an expression of x_2 in terms of x_1. If we knew x_1, this would provide a formula for us to establish the other coordinate, x_2.

Since the first equation must also be satisfied for a solution (x_1, x_2), it must be valid to replace x_2 with $-2x_1 + 1$ in that equation, which provides a path to find x_1:

$$2x_1 + (-2x_1 + 1) = 3$$

Adding the x_1 terms, we get

$$1 = 3.$$

Clearly something went wrong here, but what is it exactly? Our initial assumption was that there exists a point (x_1, x_2) that satisfies both equations, but this assumption logically implies a result that says $1 = 3$, which is clearly false.

The proof by contradiction method we learned in *Chapter 2, Formal Logic and Constructing Mathematical Proofs*, reveals that this initial assumption must have been false, so there is no such point: there is no solution to this system of equations.

In the following graph, we see that the two lines are parallel, and therefore never cross one another. This means the lines share no points, as can be seen in the following graph:

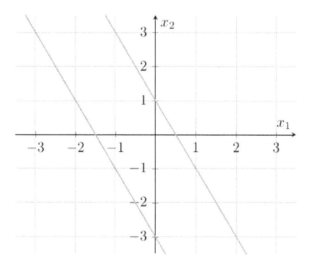

Figure 6.4 – The graphs of the two linear equations in this example are parallel, so they never cross and there are no solutions to the system

As we defined previously, a linear system of equations that are geometrically represented as parallel lines is inconsistent. This means there is no solution to the system because there is no point that touches both lines.

Example – A dependent system

Consider the following system of two linear equations:

$$2x_1 + x_2 = 1$$
$$6x_1 + 3x_2 = 3$$

To find a solution to the system, again, we need to find coordinates x_1 and x_2 such that *both* equations are satisfied simultaneously. So, suppose these coordinates exist, then we can think about what must be true about them. The second equation must be true, so if we solve it for x_2 (as we did previously), we see

$$x_2 = -2x_1 + 1,$$

an expression of x_2 in terms of x_1. If we knew x_1, this would provide a formula for us to establish the other coordinate, x_2.

Since the first equation must also be satisfied for a solution (x_1, x_2), it must be valid to replace x_2 with $-2x_1 + 1$ in that equation, which provides a path to find x_1:

$$2x_1 + (-2x_1 + 1) = 1$$

Adding the x_1 terms, we are left with only

$$1 = 1.$$

Once again, something does not seem quite right. Instead of getting the x_1 coordinate we wanted, we get a simple result that says $1 = 1$. This is obviously true, but it is not a solution to the linear system, so what does it mean?

If we take the second equation and divide it by 3, we get $2x_1 + x_2 = 1$, the same as the first equation, so the second equation is not really adding any extra information in a sense. If a point (x_1, x_2) satisfies the first equation, of course it satisfies the second, and vice versa. Therefore, each line represents the same set of infinitely many points, so any point in the form $(x_1, -2x_1 + 1)$, given a real number x_1, is a solution.

We call such systems of linear equations with infinitely many solutions dependent because we have two equations representing identical lines, as can be seen in the following graph:

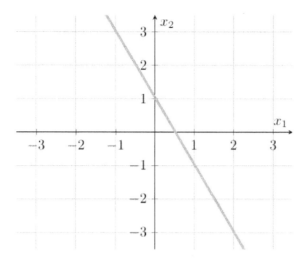

Figure 6.5 – The graphs of the two linear equations in this example are geometrically the same line, so every point on one line is on the other line, so they are all solutions to the system of equations

So far in this chapter, we have defined linear equations with two unknowns and drawn their graphs, which are lines. Then, we considered linear systems of two equations with two unknowns. As there are two linear equations, plotting the two equations results in *two* lines. Then, in a series of examples, we saw that there are three possible types of system of two linear equations:

- **Consistent system**: The lines cross at one point, which is the unique solution to the system.
- **Inconsistent system**: The lines never cross, so there are no solutions.
- **Dependent system**: The lines are the same, so all points on the line are solutions.

In the next couple of pages, we will extend these ideas to linear systems of more than two equations with more than two unknowns. Although the situation becomes more complicated in this setting, much of the preceding theory still applies. Linear systems are still classified the same way, with these same three classes.

Let's define a few more notions for these larger systems of linear equations before continuing to solve them, which turns out to be harder to do by hand than these examples, so we will turn to a number of Python functions to solve them for us once we understand the main idea of the solution method.

Definition – Systems of linear equations and their solutions

A system of n linear equations in variables $x_1, x_2, ..., x_n$ is a set of equations in the following form:

$$a_{11}x_1 + a_{12}x_2 + \cdots + a_{1n}x_n = b_1$$

$$a_{21}x_1 + a_{22}x_2 + \cdots + a_{2n}x_n = b_2$$

$$\vdots \qquad \vdots \qquad \ddots \qquad \vdots$$

$$a_{n1}x_1 + a_{n2}x_2 + \cdots + a_{nn}x_n = b_n$$

where each a_{ij} and b_i is a real constant. A solution to the system is a point $(x_1, x_2, ..., x_n)$ in n-dimensional space that solves all of the equations simultaneously.

So, just like the definition when we limited it to two equations and two variables, we seek a point that solves all the equations. However, instead of the solution being a point in a 2D coordinate plane, the solutions to these are points in a higher dimensional space. The 3D case is easy to visualize, as we are accustomed to seeing the world in three dimensions, but it is not possible to visualize the higher dimensional spaces quite so well, so the geometric interpretations of solutions are not so easy to discuss. Nevertheless, the mathematics we will present produces accurate results in those higher dimensional spaces.

Definition – Consistent, inconsistent, and dependent systems

A system of n linear equations with n variables falls into one of three categories:

- If the system has one solution, it is called consistent.
- If the system has no solutions, it is called inconsistent.
- If the system has infinitely many solutions, it is called dependent.

It is possible to solve these larger systems of linear equations by hand by means of a substitution process similar to what we did in the preceding example, but it quickly gets very long and tedious, so we will present a standard method called Gaussian elimination that always works, but it is best left to algorithms in practice. However, we need to do some pre-processing to the system to put it into a special new form with some mathematical structures called matrices before we can use Gaussian elimination (both by hand and with Python).

Matrices and matrix representations of linear systems

Solving systems of more than two equations in more than two variables is very cumbersome under the algebraic notation we used previously for the small notations, so we need an alternate notation. We will take the coefficients of a system of n linear equations with n unknowns denoted a_{ij} above and arrange them in a special sort of array called a matrix. What makes matrices distinct from arrays you may be accustomed to using in code is that matrices have a special multiplication operation that simplifies many calculations and, especially, makes solving larger linear systems much easier.

We will also represent the x_j and the b_i terms as matrices to make a single matrix equation instead of n separate equations. Once we do that, we will be ready to solve these larger systems efficiently by hand and then with Python.

Definition – Matrices and vectors

An m-by-n matrix A is a rectangular array of numbers with m rows and n columns, which have some associated mathematical operations defined between matrices and between numbers and matrices.

Each number in a matrix is called an entry or element of the matrix and the entry in the ith row and j^{th} column is typically written with a lowercase a_{ij}. A matrix is usually written in the form

$$\mathbf{A} = \begin{bmatrix} a_{11} & a_{12} & \cdots & a_{1n} \\ a_{21} & a_{22} & \cdots & a_{2n} \\ \vdots & \vdots & \ddots & \vdots \\ a_{m1} & a_{m2} & \cdots & a_{mn} \end{bmatrix} = (a_{ij})$$

Vectors are matrices with either one row or one column. The following vectors are called the column vectors of A, where each column of A will become a vector:

$$\begin{bmatrix} a_{11} \\ a_{21} \\ \vdots \\ a_{m1} \end{bmatrix}, \begin{bmatrix} a_{12} \\ a_{22} \\ \vdots \\ a_{m2} \end{bmatrix}, \dots, \begin{bmatrix} a_{1n} \\ a_{2n} \\ \vdots \\ a_{mn} \end{bmatrix}$$

The following vectors are called the column vectors of A:

$$[a_{11} \quad a_{12} \quad \cdots \quad a_{1n}], [a_{21} \quad a_{22} \quad \cdots \quad a_{2n}], \dots, [a_{m1} \quad a_{m2} \quad \cdots \quad a_{mn}]$$

In Python, we can represent the following two matrices,

$$\mathbf{A} = \begin{bmatrix} 3 & 2 & 1 \\ 9 & 0 & 1 \\ 3 & 4 & 1 \end{bmatrix} \text{ and } \mathbf{B} = \begin{bmatrix} 1 & 1 & 2 \\ 8 & 4 & 1 \\ 0 & 0 & 3 \end{bmatrix},$$

and access specific entries of the matrices in the following code:

```
import numpy

# initialize matrices
A = numpy.array([[3, 2, 1], [9, 0, 1], [3, 4, 1]])
B = numpy.array([[1, 1, 2], [8, 4, 1], [0, 0, 3]])

# print the entry in the first row and first column of A
print(A[0,0])

# print the entry in the second row and third column of B
print(B[1,2])
```

So, the code first creates the two matrices, *A* and *B*, given here. (They are called NumPy arrays in the language of Python.)

Then, we call and print the number in the very first row and very first column of matrix *A*, which in code is A[0,0], but in mathematical notation is $a_{11} = 3$, which the code outputs. Lastly, we similarly choose B[1,2], the element in row 2 and column 3 of matrix *B*, in other words, $b_{23} = 1$

```
3
1
```

It is important to be aware that Python and most other programming languages begin indexing arrays (and matrices) with 0 while mathematicians tend to start with 1, which is why the numbers in the code are one less than the mathematical language would suggest.

> **Important note**
> There is a matrix class built into NumPy that has been used for linear algebra, but current documentation says this class will be deprecated in the future, so users should use arrays instead. We will follow this convention.

Now that we have some common vocabulary about matrices, we will discuss ways to manipulate matrices, multiply them with numbers, add and subtract matrices, and multiply matrices. These operations are what distinguish matrices from ordinary arrays.

Definition – Matrix addition and subtraction

Let $A = (a_{ij})$ and $B = (b_{ij})$ be *m*-by-*n* matrices. Their sum is found by simply adding the entries of each matrix elementwise, meaning each a_{ij} is added to each b_{ij} as follows:

$$A + B = \begin{bmatrix} a_{11} + b_{11} & a_{12} + b_{12} & \cdots & a_{1n} + b_{1n} \\ a_{21} + b_{21} & a_{22} + b_{22} & \cdots & a_{2n} + b_{2n} \\ \vdots & \vdots & \ddots & \vdots \\ a_{m1} + b_{m1} & a_{m2} + b_{m2} & \cdots & a_{mn} + b_{mn} \end{bmatrix}$$

In other words, we add up the terms in the same positions in matrix *A* and in matrix *B*.

And the difference in two matrices works similarly, as can be seen here:

$$A - B = \begin{bmatrix} a_{11} - b_{11} & a_{12} - b_{12} & \cdots & a_{1n} - b_{1n} \\ a_{21} - b_{21} & a_{22} - b_{22} & \cdots & a_{2n} - b_{2n} \\ \vdots & \vdots & \ddots & \vdots \\ a_{m1} - b_{m1} & a_{m2} - b_{m2} & \cdots & a_{mn} - b_{mn} \end{bmatrix}$$

In other words, we subtract up the terms in the same positions in matrix *A* and in matrix *B*.

> **Important note**
>
> The sum and difference of two matrices is only defined if the two matrices have the same dimensions, in other words, the same number of rows and the same number of columns.

We can use the `numpy.add` and `numpy.subtract` functions to add and subtract matrices in Python as in the following code, which follows from the preceding code:

```
# Add A and B
print(numpy.add(A,B))

# Subtract A and B
print(numpy.subtract(A,B))
```

The code has the following output:

```
[[ 4   3   3]
 [17   4   2]
 [ 3   4   4]]

[[ 2   1  -1]
 [ 1  -4   0]
 [ 3   4  -2]]
```

Of course, this is in fact $A + B$ and $A - B$, which we could find by hand if we add and subtract all the numbers in the matrices element by element.

Next, we continue with more arithmetic of matrices: multiplying a whole matrix by a scalar (or, by a number).

Definition – Scalar multiplication

Let $c \in R$ be a real number. Such a constant is frequently referred to as a scalar. The product of this scalar c and a matrix A is defined as a matrix where each element is the product of c times the corresponding element of A:

$$c\mathbf{A} = c \begin{bmatrix} a_{11} & a_{12} & \cdots & a_{1n} \\ a_{21} & a_{22} & \cdots & a_{2n} \\ \vdots & \vdots & \ddots & \vdots \\ a_{m1} & a_{m2} & \cdots & a_{mn} \end{bmatrix} = \begin{bmatrix} ca_{11} & ca_{12} & \cdots & ca_{1n} \\ ca_{21} & ca_{22} & \cdots & ca_{2n} \\ \vdots & \vdots & \ddots & \vdots \\ ca_{m1} & ca_{m2} & \cdots & ca_{mn} \end{bmatrix}$$

In simpler terms, we simply take our real number c and multiply it by each and every number in the matrix.

As we see, the sum and differences of matrices and the scalar multiplication of matrices are somewhat obvious, as we simply do the operations for each element. Matrix multiplication, on the other hand, is not simply elementwise multiplication.

Before that, we define transposes of matrices and a special case of the matrix product called the dot product, which is limited to multiplying a row vector by a column vector.

Definition – Transpose of a matrix

Let $A = (a_{ij})$ be an *m*-by-*n* matrix. The transpose of *A*, denoted *AT*, is the *n*-by-*m* matrix resulting from switching each element in the ith row and j^{th} column of *A* to the element in the *jth* row and *ith* column of the new matrix,

$$\mathbf{A}^T = \begin{bmatrix} a_{11} & a_{21} & \cdots & a_{n1} \\ a_{12} & a_{22} & \cdots & a_{n2} \\ \vdots & \vdots & \ddots & \vdots \\ a_{1m} & a_{2m} & \cdots & a_{nm} \end{bmatrix}.$$

In simpler terms, a transpose moves the elements of a matrix around by swapping the row of an element with its column. Here are a couple of examples:

- Element a_{21} in row 2, column 1 of the original matrix *A* moves to row 1, column 2 in the new transpose matrix, A^T.
- Element a_{n1} in row n, column 1 of the original matrix *A* moves to row 1, column *n* in the new transpose matrix, A^T.

> **Important note**
>
> The transpose of a matrix in general has different dimensions to the original matrix, with the number of rows and the number of columns interchanged.

We can also use NumPy to do scalar multiplication and find transposes, as the following code, continuing on from the previous code, will do:

```
# Multiply A by a scalar 5
print(numpy.multiply(5,A))

# Find the transpose of A
print(numpy.transpose(A))
```

The output of this code is as expected:

```
[[15 10  5]
 [45  0  5]
 [15 20  5]]

[[3 9 3]
 [2 0 4]
 [1 1 1]]
```

The first multiplication of 5 with the matrix A multiplies each element of the original matrix by the number 5. The second part takes a transpose properly by swapping the rows with the columns of the original A matrix.

To wrap up the section, we will look at multiplication not between a number and a matrix, but multiplication between two matrices, which has some special rules. This allows us to convert systems of linear equations of any size into a single matrix equation.

Definition – Dot product of vectors

The dot product of a 1-by-n row vector a and an n-by-1 column vector b is defined as

$$\mathbf{a} \cdot \mathbf{b}^T = \mathbf{a}\mathbf{b}^T = \begin{bmatrix} a_1 & a_2 & \cdots & a_n \end{bmatrix} \begin{bmatrix} b_1 \\ b_2 \\ \vdots \\ b_n \end{bmatrix}^T = \sum_{j=1}^{n} a_{1j} b_{j1} = a_{11} b_{11} + a_{12} b_{21} + \cdots + a_{1n} b_{n1}$$

In other words, we multiply the first number in a by the first number in b, the second number in a by the second number in b, and so on, and add up all of the results of these multiplications.

In general, matrix multiplication for larger matrices computes dot products of the rows of the first matrix and columns of the second matrix.

Definition – Matrix multiplication

Let A be an n-by-m and let B be an m-by-p matrix, written in the forms

$$\mathbf{A} = \begin{bmatrix} \mathbf{a}_1 \\ \mathbf{a}_2 \\ \vdots \\ \mathbf{a}_n \end{bmatrix} \text{ and } \mathbf{B} = \begin{bmatrix} \mathbf{b}_1^T & \mathbf{b}_2^T & \cdots & \mathbf{b}_p^T \end{bmatrix},$$

where each a_i and b_j is a 1-by-m column vector. So, we represent our matrix A by stacking up its horizontal rows $a_1, a_2, \ldots, a_n$, and we represent our matrix B by stacking its vertical columns side by side.

The product of the matrices is denoted by AB and the element of AB in the ith row and j^{th} column is the dot product of the ith row of A and the jth column of B, as follows:

$$\mathbf{AB} = \begin{bmatrix} \mathbf{a}_1 \mathbf{b}_1^T & \mathbf{a}_1 \mathbf{b}_2^T & \cdots & \mathbf{a}_1 \mathbf{b}_p^T \\ \mathbf{a}_2 \mathbf{b}_1^T & \mathbf{a}_2 \mathbf{b}_2^T & \cdots & \mathbf{a}_2 \mathbf{b}_p^T \\ \vdots & \vdots & \ddots & \vdots \\ \mathbf{a}_n \mathbf{b}_1^T & \mathbf{a}_n \mathbf{b}_2^T & \cdots & \mathbf{a}_n \mathbf{b}_p^T \end{bmatrix}$$

In simpler terms, matrix multiplication takes the dot product of each row vector of A with each column vector of B.

> **Important note**
> The matrix product AB is only defined when A has the same number of columns as B has rows, and AB has the same number of rows as A and the same number of columns as B. Thus, multiplying an n-by-m matrix by an m-by-p matrix is permitted and results in an n-by-p matrix.

This definition can feel a little difficult, so next, we will do an example where we carefully multiply two matrices by hand and then do it in Python.

Example – Multiplying matrices by hand and with NumPy

Define two matrices as

$$\mathbf{A} = \begin{bmatrix} 1 & 3 & 1 \\ 2 & 3 & 0 \end{bmatrix} \text{ and } \mathbf{B} = \begin{bmatrix} 3 & 5 \\ 1 & 0 \\ 2 & 2 \end{bmatrix}.$$

Then, the product can be computed as

$$\mathbf{AB} = \begin{bmatrix} 1 & 3 & 1 \\ 2 & 3 & 0 \end{bmatrix} \begin{bmatrix} 3 & 5 \\ 1 & 0 \\ 2 & 2 \end{bmatrix}.$$

By the definition of matrix multiplication, this is a matrix made up of dot products of each row of A with each column of B:

$$\mathbf{AB} = \begin{bmatrix} \mathbf{a}_1 \mathbf{b}_1^T & \mathbf{a}_1 \mathbf{b}_2^T \\ \mathbf{a}_2 \mathbf{b}_1^T & \mathbf{a}_2 \mathbf{b}_2^T \end{bmatrix}$$

To do each of the four dot products, we multiply the first term of a row by the first term of a column, the second term of a row by the second term of a column, and finally the third term in a row by the third term of a column and add them up:

$$\mathbf{AB} = \begin{bmatrix} (1)(3) + (3)(1) + (1)(2) & (1)(5) + (3)(0) + (1)(2) \\ (2)(3) + (3)(1) + (0)(2) & (2)(5) + (3)(0) + (0)(2) \end{bmatrix}$$

Simplifying the arithmetic, we get

$$AB = \begin{bmatrix} 8 & 7 \\ 9 & 10 \end{bmatrix}$$

Suppose we were to multiply *BA* instead. This is possible since *B* is 3-by-2 and *A* is 2-by-3, so the result *BA* will be a 3-by-3 matrix. Unlike ordinary multiplication, with matrix multiplication, we have *AB* ≠ *BA* in some cases. Mathematically, this means matrix multiplication is not a commutative operation: the order of the factors matters.

We see matrix multiplication is easy enough to do by hand for a small matrix, but if the matrices were much larger, the number of steps in the arithmetic would make this pretty inefficient, so we generally prefer to use code. We may multiply matrices with NumPy with the following code. Again, this continues on from the previous code in this section:

```
# Multiply A and B
print(numpy.dot(A,B))
```

The output is as follows

```
[[19 11 11]
 [ 9  9 21]
 [35 19 13]]
```

> **Important note**
>
> NumPy has `numpy.multiply` and `numpy.dot` functions.
>
> `numpy.multiply` performs component-wise multiplication.
>
> `numpy.dot` performs matrix multiplication, as we defined previously.

Now that we know about matrix multiplication, the definition of a system of *n* linear equations with *n* unknowns $x_1, x_2, ..., x_n$ can be written in terms of matrix multiplication as

$$\begin{bmatrix} a_{11} & a_{12} & \cdots & a_{1n} \\ a_{21} & a_{22} & \cdots & a_{2n} \\ \vdots & \vdots & \ddots & \vdots \\ a_{n1} & a_{n2} & \cdots & a_{nn} \end{bmatrix} \begin{bmatrix} x_1 \\ x_2 \\ \vdots \\ x_n \end{bmatrix} = \begin{bmatrix} b_1 \\ b_2 \\ \vdots \\ b_n \end{bmatrix},$$

which can be written much more compactly as *Ax* = *b* and we call it an *n*-by-*n* linear system, corresponding to the dimensions of *A*. Now, to see why this is true if things aren't clear, let's multiply out the matrix and see what happens. If we compute the dot product of row 1 of *A* by the *x* vector and set it equal to the top number in the *b* vector, we get

$$a_{11}x_1 + a_{12}x_2 + \cdots + a_{1n}x_n = b_1,$$

which is exactly the first equation in our system!

For another one, let's multiply the dot product row 2 of A by the x vector and set it equal to the second number in b:

$$a_{21}x_1 + a_{22}x_2 + \cdots + a_{2n}x_n = b_2,$$

which is the second equation in the system. Continuing this for each row, the matrix multiplication of A by x will generate each equation, one by one. So, we see this sort of matrix equation is equivalent to all n equations in our system, with each row corresponding to one of the equations.

Another common representation that we will use is a so-called augmented matrix to represent the system as follows:

$$[\mathbf{A}|\mathbf{b}] = \begin{bmatrix} a_{11} & a_{21} & \cdots & a_{n1} & b_1 \\ a_{12} & a_{22} & \cdots & a_{n2} & b_2 \\ \vdots & \vdots & \ddots & \vdots & \vdots \\ a_{1m} & a_{2m} & \cdots & a_{nm} & b_n \end{bmatrix}.$$

Each row in the augmented matrix *[A|b]* corresponds to one of the equations. The ith row of *[A|b]* corresponds to the ith equation of the system:

$$a_{i1}x_1 + a_{i2}x_2 + \ldots + a_{in}x_n = b_i$$

Next, we look at an approach to solve these linear systems called Gaussian elimination and learn how to implement it with code.

Solving small linear systems with Gaussian elimination

In this section, we will learn how to solve an *n*-by-*n* linear system of equations $Ax = b$, if possible, through a method called **Gaussian elimination**, which we will do by hand for a small problem. In the next section, we implement it with Python.

We will explain through an example of a 3-by-3 system, which should make the idea clear for larger systems, which we will formalize at the end of the section, and which we will prefer to solve with code.

First, notice that there are several manipulations we may do to the equation in the system without changing the solutions:

- We can switch the order of the equations, which corresponds to swapping the rows of the matrix *[A|b]*.
- We can multiply both sides of an equation by a constant, which corresponds to multiplying a row of *[A|b]* by a constant.
- We can add a multiple of one equation to another equation, which corresponds to adding a multiple of one row of *[A|b]* to another row.

The effects of the augmented matrix corresponding to each of these manipulations are called elementary row operations. Gaussian elimination is a method that will use specific sequences of row operations to manipulate the system into a very simple version that will make it immediately solvable or reveal it to be inconsistent or dependent. The form is shown next.

Definition – Leading coefficient (pivot)

For each row of a matrix not fully filled with zeros, the leading coefficient (or pivot) of the row is the first non-zero number in the row.

For example, consider the matrix

$$\begin{bmatrix} 2 & 3 & 0 \\ 0 & 1 & 2 \\ 0 & 5 & 4 \end{bmatrix}.$$

The pivots of this matrix are the 2 in the first row, the 1 in the second row, and the 5 in the third row.

Definition – Reduced row echelon form

A matrix is in **reduced row echelon form** (**RREF**) if the following applies:

- Any zero rows are on the bottom.
- The pivot of each non-zero row is a 1 and is to the right of the pivot of the previous row.
- Each column containing a 1 pivot has zeros in all other entries.

Example – Consistent system in RREF

For example, the following matrix is in RREF:

$$\begin{bmatrix} 1 & 0 & 0 & | & 2 \\ 0 & 1 & 0 & | & 3 \\ 0 & 0 & 1 & | & 1 \end{bmatrix}$$

The corresponding linear system is simply

$$1x_1 + 0x_2 + 0x_3 = 2$$
$$0x_1 + 1x_2 + 0x_3 = 3$$
$$0x_1 + 0x_2 + 1x_3 = 1.$$

In other words, an augmented matrix in RREF immediately gives the solution of a linear system, *(2, 3, 1)* in this case, if it has a pivot in each row and we know the system is, therefore, consistent. In most cases, we will have a more complex augmented matrix initially that is transformed into this form using elementary row operations.

Example – Inconsistent system in RREF

Suppose a system has an augmented matrix in RREF as follows:

$$\begin{bmatrix} 1 & 0 & 0 & | & 4 \\ 0 & 1 & 1 & | & 1 \\ 0 & 0 & 0 & | & 2 \end{bmatrix}$$

Even though the third column has multiple numbers in it, this is permitted since that column does not contain a pivot 1. The corresponding system of equations is

$$1x_1 + 0x_2 + 0x_3 = 4$$
$$0x_1 + 1x_2 + 1x_3 = 1$$
$$0x_1 + 0x_2 + 0x_3 = 2.$$

However, the third row suggests *0 = 2*, a contradictory statement. Just like the case with a system of two linear equations, this means the system is inconsistent – in other words, it has no solutions. In 2D, this means the lines are parallel, but in 3D, the equations represent planes, so it means at least two of the planes are parallel.

Example – Dependent system in RREF

Lastly, consider a system with the following RREF augmented matrix:

$$\begin{bmatrix} 1 & 0 & 1 & | & 1 \\ 0 & 1 & 2 & | & 6 \\ 0 & 0 & 0 & | & 0 \end{bmatrix}$$

If we convert this RREF form back into equation form, we have

$$1x_1 + 0x_2 + 1x_3 = 1$$

$$0x_1 + 1x_2 + 2x_3 = 6$$

$$0x_1 + 0x_2 + 0x_3 = 0.$$

The last line is true but tells us nothing about x_3. Just like the 2D case, this means the system is dependent. In this situation, x_3 is called a free variable because we can construct a point solving the system for any given x_3 value.

Given x_3, we know that $x_1 = 1 - x_3$ and $x_2 = 6 - 2x_3$, so the solution set for this system of three linear equations is $\{(1 - x_3, 6 - 2x_3, x_3) : x_3 \in R\}$, in other words, any ordered pair in this form is a solution.

Now we have learned the three permissible row operations and that we can easily determine the solution of a linear system if we have transformed the system's augmented matrix into RREF, so the question becomes, simply: Which row operations do we need to do to go from a given augmented matrix to RREF?

Gaussian elimination answers this question by providing a sequence of row operations that will never fail to do this conversion, which we define now.

Algorithm – Gaussian elimination

The specifics of different implementations of this method may vary and have some changes to optimize the calculations, but the most direct approach is provided by the following pseudocode:

Step 1: Re-order the rows of $[A|b]$ from i to n so that the leftmost pivot is and pivots in subsequent rows are in the same column or to the right of the pivot of the previous row.

Step 2: Set $i = 1$.

Step 3: Divide row i by its pivot.

Step 4: Add multiples of row i to each successive row chosen such that the numbers under the pivot of row i become zeros.

Step 5: Add *1* to *i*.

Step 6: Move all zero rows of *[A|b]* to the bottom. Set *m* to be the number of zero rows.

Step 7: If $i < n - m$, return to *Step 3*. Otherwise, $i = n - m$ and continue to *Step 8*.

Step 8: If row *i* has a pivot, add multiples of row *i* to all previous rows chosen such that the numbers above the pivot of row *i* become 0. Otherwise, proceed immediately to *Step 9*.

Step 9: Subtract *1* from *i*.

Step 10: If $i = 1$, terminate. Otherwise, return to *Step 8*.

The first phase of Gaussian elimination (*Steps 1-6*) ensures the matrix has all pivots set to 1 with 0s under them. The second phase (*Steps 8-10*) fills in 0s above the pivots so that the matrix is converted to RREF. Augmented matrices will represent the same linear system as the original *[A|b]*.

Example – 3-by-3 linear system

Consider the system of linear equations:

$$2x_1 - 6x_2 + 6x_3 = -8$$

$$2x_1 + 3x_2 - x_3 = 15$$

$$4x_1 - 3x_2 - x_3 = 10$$

We can write this system of equations as the following augmented matrix:

$$\begin{bmatrix} 2 & -6 & 6 & | & -8 \\ 2 & 3 & -1 & | & 15 \\ 4 & -3 & -1 & | & 19 \end{bmatrix}$$

First, divide row 1 by 2 to get the following:

$$\begin{bmatrix} 2 & -6 & 6 & | & -8 \\ 2 & 3 & -1 & | & 15 \\ 4 & -3 & -1 & | & 19 \end{bmatrix} \xrightarrow{\frac{1}{2}(\text{row 1})} \begin{bmatrix} 1 & -3 & 3 & | & -4 \\ 2 & 3 & -1 & | & 15 \\ 4 & -3 & -1 & | & 19 \end{bmatrix}$$

Next, add -2 times the first row to row 2 and -4 times the first row to row 3 to fill in zeros under the first pivot:

$$\begin{bmatrix} 1 & -3 & 3 & | & -4 \\ 2 & 3 & -1 & | & 15 \\ 4 & -3 & -1 & | & 19 \end{bmatrix} \xrightarrow{\text{row 2}+(-2)(\text{row 1})} \begin{bmatrix} 1 & -3 & 3 & | & -4 \\ 0 & 9 & -7 & | & 23 \\ 4 & -3 & -1 & | & 19 \end{bmatrix} \xrightarrow{\text{row 3}+(-4)(\text{row 1})} \begin{bmatrix} 1 & -3 & 3 & | & -4 \\ 0 & 9 & -7 & | & 23 \\ 0 & 9 & -13 & | & 35 \end{bmatrix}$$

Then, divide row 2 by 9 so its pivot becomes 1:

$$\begin{bmatrix} 1 & -3 & 3 & | & -4 \\ 0 & 9 & -7 & | & 23 \\ 0 & 9 & -13 & | & 35 \end{bmatrix} \xrightarrow{\frac{1}{9}(\text{row 2})} \begin{bmatrix} 1 & -3 & 3 & | & -4 \\ 0 & 1 & -\frac{7}{9} & | & \frac{23}{9} \\ 0 & 9 & -13 & | & 35 \end{bmatrix}$$

Add -9 times row 2 to row 3:

$$\begin{bmatrix} 1 & -3 & 3 & | & -4 \\ 0 & 1 & -\frac{7}{9} & | & \frac{23}{9} \\ 0 & 9 & -13 & | & 35 \end{bmatrix} \xrightarrow{\text{row 3}+(-9)(\text{row 2})} \begin{bmatrix} 1 & -3 & 3 & | & -4 \\ 0 & 1 & -\frac{7}{9} & | & \frac{23}{9} \\ 0 & 0 & -6 & | & 12 \end{bmatrix}$$

Dividing row 3 by -6 completes the first phase of Gaussian elimination:

$$\begin{bmatrix} 1 & -3 & 3 & | & -4 \\ 0 & 1 & -\frac{7}{9} & | & \frac{23}{9} \\ 0 & 0 & -6 & | & 12 \end{bmatrix} \xrightarrow{\frac{1}{-6}(\text{row 3})} \begin{bmatrix} 1 & -3 & 3 & | & -4 \\ 0 & 1 & -\frac{7}{9} & | & \frac{23}{9} \\ 0 & 0 & 1 & | & -2 \end{bmatrix}$$

The remaining steps will comprise the second phase of Gaussian elimination. To fill in zeros above the last pivot, add 7/9 times row 3 to row 2 and add -3 times row 3 to row 1:

$$\begin{bmatrix} 1 & -3 & 3 & | & -4 \\ 0 & 1 & -\frac{7}{9} & | & \frac{23}{9} \\ 0 & 0 & 1 & | & -2 \end{bmatrix} \xrightarrow{\text{row 2}+\frac{7}{9}(\text{row 3})} \begin{bmatrix} 1 & -3 & 3 & | & -4 \\ 0 & 1 & 0 & | & 1 \\ 0 & 0 & 1 & | & -2 \end{bmatrix} \xrightarrow{\text{row 1}+(-3)(\text{row 3})} \begin{bmatrix} 1 & -3 & 0 & | & 2 \\ 0 & 1 & 0 & | & 1 \\ 0 & 0 & 1 & | & -2 \end{bmatrix}$$

Lastly, add 3 times row 2 to row 1 to get to the RREF:

$$\begin{bmatrix} 1 & -3 & 0 & | & 2 \\ 0 & 1 & 0 & | & 1 \\ 0 & 0 & 1 & | & -2 \end{bmatrix} \xrightarrow{\text{row 1}+(3)(\text{row 2})} \begin{bmatrix} 1 & 0 & 0 & | & 5 \\ 0 & 1 & 0 & | & 1 \\ 0 & 0 & 1 & | & -2 \end{bmatrix}$$

Thus, this RREF form reveals the system is consistent and its solution *(5, 1, -2)*, is the numbers in the rightmost column.

In this section, we have introduced the reduced row echelon form (RREF) of a matrix corresponding to a linear system of equations introduced in the previous sections. The RREF always easily reveals the solution to the system if it exists, or reveals the system is inconsistent if there is no solution. After that, we considered an algorithm called Gaussian elimination that never to convert the matrix corresponding to an *n*-by-*n* linear system of equations to the RREF and, therefore, reveals the solution if possible. Lastly, we applied the algorithm by hand to a small, 3-by-3 linear system.

Now that we understand the problem Gaussian elimination solves and how it works, we will continue to learn how to implement the algorithm with NumPy.

Solving large linear systems with NumPy

The last example should make it clear that Gaussian elimination will work for any linear system to reduce it to RREF form, but this 3-equation, 3-variable system required a significant amount of effort to solve, and things will only become more complex for larger systems, so the more practical way to do it is to use existing algorithms. In this section, we will learn how to use some methods with NumPy to accomplish this task.

A Python function for solving systems of linear equations $Ax = b$ is available in NumPy named `numpy.linalg.solve`, which works for square, consistent systems. That is, it finds solutions for all linear systems that have unique solutions.

Typically, the function uses a version of Gaussian elimination just as we have done by hand, but it is a very smart function. First, the function chooses the order of calculations carefully to optimize its speed. Second, if the function detects that A has a special structure (such as a symmetric, diagonal, or banded matrix), it will take shortcuts and use variants of Gaussian elimination and other methods that exploit the structure to run even faster!

Although we will not delve into these other methods since it would require an in-depth study of linear algebra, we can think of the function in terms of what it accomplishes and avoid the details. Nevertheless, we benefit from the speed-ups.

Let's try it!

Example – A 3-by-3 linear system (with NumPy)

We begin with a system we already solved by hand just to be sure we get the same answer from the NumPy function, so consider the linear system $Ax = b$ with the augmented matrix form:

$$[\mathbf{A}|\mathbf{b}] = \begin{bmatrix} 2 & -6 & 6 & | & -8 \\ 2 & 3 & -1 & | & 15 \\ 4 & -3 & -1 & | & 19 \end{bmatrix}$$

To solve this, we need to create two NumPy matrices, one for A and one for b to feed into the `numpy.linalg.solve` function and then run it:

```
import numpy

# Create A and b matrices
A = numpy.array([[2, -6, 6], [2, 3, -1], [4, -3, -1]])
b = numpy.array([-8, 15, 19])

# Solve Ax = b
numpy.linalg.solve(A,b)
```

The `numpy.linalg.solve(A,b)` line runs an optimized version of Gaussian elimination and code returns:

```
array([ 5.,  1., -2.])
```

In other words, the code tells us the solution is (5, 1, -2), just as we found by hand, but this time we get the solution almost instantaneously!

Example – Inconsistent and dependent systems with NumPy

We said `numpy.linalg.solve` requires consistent systems, but a reasonable question is what happens if you give it matrices A and b corresponding to an inconsistent or dependent system, so let's try the inconsistent and dependent systems we considered in the first example of the chapter.

In the following code, we repeat the same idea as the previous example twice more, but this time, we solve the inconsistent and dependent problems we solved before:

```
import numpy

# inconsistent system
A = numpy.array([[2, 1], [6, 3]])
b = numpy.array([3, 3])

print(numpy.linalg.solve(A,b))

# dependent system
A = numpy.array([[2, 1], [6, 3]])
b = numpy.array([1, 3])

print(numpy.linalg.solve(A,b))
```

The output is as follows:

```
[-1.80143985e+16  3.60287970e+16]
[0. 1.]
```

For the inconsistent system, it returns some giant numbers in the order of *10*^{16}, but we know there is no solution mathematically, so this is meaningless. In the case of the dependent system, it returns (0, 1), which is a solution to the system, but it has infinitely many solutions.

A key take-away from this example is that we should never implement numpy.linalg.solve without carefully screening the coefficient matrix A we will feed into it because it will return nonsense or incomplete answers *without* giving us any sort of warning.

How can we test A? The theory required is beyond the scope of this book, but there is a number called a determinant that can be computed for a square matrix and there is a theorem called the Invertible Matrix Theorem that tells us a linear system is always consistent if the determinant of A is not 0. Therefore, a good practice is to verify that the determinant of A is nonzero with the numpy.linalg.det function before proceeding further. In the following code, we create a NumPy array A, compute the determinant, and print it:

```
A = numpy.array([[2, 1], [6, 3]])
print(numpy.linalg.det(A))
```

This produces the following output:

```
-3.330669073875464e-16
```

This is a number in the order of 10^{-16}, which is extremely tiny and suggests the determinant is effectively 0, which would indicate numpy.linalg.solve should not be used. In a practical implementation, we would want to check this first and output an error if the determinant is within 10^{-5} of 0 in order to account for rounding errors.

Where the numpy.linalg.solve method really shines is in larger linear systems of equations, which would be totally unreasonable to do by hand. The 3-by-3 linear system took a whole 2 pages to solve with Gaussian elimination. It would be almost unthinkable to solve a system of 10, 20, or 100 equations by hand! It turns out, however, that these are not difficult for NumPy.

Example – A 10-by-10 linear system (with NumPy)

It would be cumbersome to even write down a 10-by-10 linear system of equations, so we will rely on the short expression $Ax = b$, but which system will we solve? To save the trouble of making up a problem, suppose we just generate matrices A and b with uniformly random numbers.

It requires some deep mathematics beyond the scope of this book, but it can be shown such a system is consistent with probability 1, so it is highly likely to satisfy the requirements for numpy.linalg.solve():

The following code will do three primary things:

- Generate A and b as 10-by-10 and 10-by-1 matrices, respectively, with elements selected uniformly at random from the interval *[-5, 5]* using numpy.random.rand.
- Use numpy.linalg.solve to find a 10-by-1 matrix *x* that solves the system, and then calculates *Ax - b* using numpy.dot for the multiplication.
- Sum the absolute values of the 10-by-1 matrix *Ax - b* to verify the result is nearly *0*, possibly with some tiny error resulting from rounding errors, to confirm the solution is correct:

```
import numpy

numpy.random.seed(1)

# Create A and b matrices with random
A = 10*numpy.random.rand(10,10)-5
b = 10*numpy.random.rand(10)-5

# Solve Ax = b
solution = numpy.linalg.solve(A,b)
print(solution)

# To verify the solution works, show Ax - b is near 0
sum(abs(numpy.dot(A,solution) - b))
```

This returns the following output:

```
[ 0.09118027 -1.51451319 -2.48186344 -2.94076307
  0.07912968  2.76425416  2.48851394 -0.30974375
 -1.97943716  0.75619575]

1.1546319456101628e-14
```

As you can see, numpy.linalg.solve(A,b) finds that the solution to this 10-by-10 system of linear equations is roughly *(0.09, -1.51, -2.48, -2.94, 0.08, 2.76, 2.49, -0.31, -1.98, 0.76)*.

Note that sum and abs are some common Python function for adding and finding the absolute values of elements of a matrix. The last line of code sums the absolute values of *Ax - b* to get a number of nearly *0*. This tells us *Ax - b* is approximately *0*, so we have confirmation that the solution is accurate, at least to 14 decimal places.

> **Important note**
>
> The numpy.random.seed(1) command is used simply so that the code is reproducible for readers.
>
> If you remove this line, the calls to numpy.random.rand will generate different random numbers so that it generates a different 10-by-10 linear system of equations and solves it.
>
> This is a good practice for testing code with randomness.

Once again, this code runs almost instantaneously on a typical laptop, even though it solves a rather large system by the standards of what we can compute by hand. In fact, the performance of this function is truly extraordinary: it runs almost instantaneously even if we replace a 10-by-10 system with a 1,000-by-1,000 system. Indeed, it solves a linear system of 1,000 equations with 1,000 variables with sums of absolute errors in the order of *10-10* in almost no time!

Summary

In this chapter, we covered a lot of ground! We began by taking the familiar idea of a linear equation in two variables and demonstrated that the set of points that satisfy the equation are exactly those that form a straight line. We then extended this to a *system* of two linear equations of two variables, which represent, geometrically speaking, two lines. A solution to the system is a point that satisfies not one, but both equations. Geometrically, this means a solution can only be a point of intersection of the two lines. As we know from elementary geometry, two lines must either be parallel, intersect, or coincide entirely. This characterizes three possible conclusions about solutions: a system must have no solutions (if they are parallel), one unique solution (if they intersect), or infinitely many solutions (if they coincide).

Then, the real fun started as we introduced systems of many linear equations and many unknowns, which are not so easily interpretable from a geometrical perspective, but they share this same property, that there must be zero, one, or infinitely many solutions.

We found these larger problems to be very cumbersome to solve with basic algebra, so we introduced a new mathematical structure to tackle this problem more efficiently: the matrix. We learned matrices behave similar to an ordinary rectangular array of numbers, but there is a special notion of multiplication of matrices that allows us to represent even large linear systems of equations in a compact, rectangular form that makes computation efficient.

After learning how to perform these matrix operations using NumPy functions, we proceeded to solve larger linear systems of equations. We learned an efficient method called Gaussian elimination that solves these systems, solved an example system of three linear equations with three variables, and then proceeded to learn to use NumPy's implementation of optimized Gaussian elimination in Python.

As we proceed through the remaining chapters, we will actually be using these methods very frequently, as linear algebra is critical for a wide array of applied problems that we will study, ranging from problems on trees and networks, cryptography, and regression analysis, to image processing and principal component analysis.

The next chapter is the first in a sequence of chapters based on tree, graph, and network structures, which are incredibly important for modeling many things, such as decision trees that guide helpdesk workers through best practices, linking structures of the web, and computer networks. In this next chapter, we define these structures, learn what they commonly model, use Python to store them efficiently, learn how to use linear algebra to find some features of the structures, and examine some key mathematical results in the area of graph theory with implications for computer science.

7
Computational Requirements for Algorithms

Algorithms that solve useful problems are at the heart of computer science, but an algorithm must not only be proven to work to be practical. They may take too long to run with our computational resources, or it may require storage of more data than our resources allow. This chapter is dedicated to finding the amount of time and space required to run algorithms; in short, the computational complexity of algorithms when it comes to time and space requirements to run a certain algorithm. We will focus on the complexity of foundational control structures and popular exemplar algorithms of common classes of time and space complexity. Different algorithms will be implemented using Python and the trade-off when it comes to runtime, computational resources, and suchlike will be discussed.

By the end of this chapter, you should have learned about different algorithms, their computational complexities, runtime, and the space required.

In this chapter, we will be covering the following topics:

- Computational complexity of algorithms
- Complexity of algorithms with fundamental control structures
- Complexity of common search algorithms
- Common classes of computational complexity

> **Important Note**
> Please navigate to the graphic bundle link to refer to the color images for this chapter.

Computational complexity of algorithms

In this section, we will learn about what algorithms are, the complexity of algorithms, and what they mean in terms of time and space and Big-O notation (compact notation for classifying the time and space needed for an algorithm). By the end of this section, you should have a good understanding of what algorithms are and their characteristics, such as complexity, and be able to determine the Big-O notation for the complexity of algorithms.

Algorithms are a step-by-step procedure/instruction to solve a problem or to obtain a desired output. They can be implemented in any programming language. Some of the important categories of algorithms from a data structure point of view are as follows:

- **Search**: Used to search for an item in a data structure
- **Sort**: Used to sort items in a required order
- **Insert**: Used to insert items into a data structure
- **Update**: Used to update an existing item in a data structure
- **Delete**: Used to delete an existing item in a data structure

Let's try out some of these algorithms in Python. We will perform the following tasks, for example:

Step 1: Ask the user to input the name of their favorite fruit.

Step 2: Append the user input fruit name to a pre-existing list of fruit names.

Step 3: Update the list and display the new list.

Step 4: Now we will delete a selected element from the list; the user inputs the name of the fruit they want to delete (Note: the name entered is case sensitive).

Step 5: Update the list and display the new list:

```python
#Type of algorithm - inserting new element to pre-existing list

fruit_name = ["Jackfruit", "Honeydew", "Grapes"]
user_input1 = input("Please enter a fruit name: ")
fruit_name.append(user_input1)
print('The updated list is: ' + str(fruit_name))


#Type of algorithm - deleting element from list

user_input2 = input("Please enter the name of the fruit you want to delete: ")
fruit_name.remove(user_input2)
print('The updated list is: ' + str(fruit_name))
```

Output:

```
Please enter a fruit name: Apple
The updated list is: ['Jackfruit', 'Honeydew', 'Grapes', 'Apple']
Please enter the name of the fruit you want to delete: Apple
The updated list is: ['Jackfruit', 'Honeydew', 'Grapes']

Process finished with exit code 0
```

In the preceding example, we have learned how to write an algorithm and display the desired results. We first added the name of a fruit of our choice to the `fruit_name` list, and then deleted a fruit name from the same list.

An algorithm must satisfy the following criteria in order to be called an algorithm:

- **Input**: It should have zero or more well-defined inputs.
- **Output**: It should have one or more well-defined outputs; often, this is the desired end product of the algorithm.
- **Finiteness**: It should terminate after a finite number of steps. While using loops, it should be ensured that the algorithm either ends after a certain number of steps or when the desired output is achieved (in a finite number of steps).

142   Computational Requirements for Algorithms

- **Feasibility**: It should be feasible with the computing resources available.
- **Unambiguous**: The instructions should be clear and have only one meaning with clearly defined inputs and outputs.

The following diagram shows how a problem can be solved by making use of multiple algorithms. However, the ideal way to solve any problem will be to choose the most efficient algorithm:

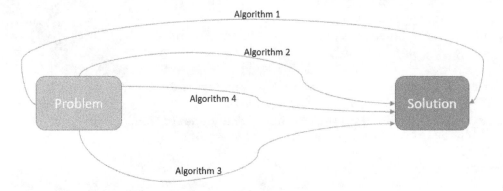

Figure 7.1 – Multiple algorithms to solve a problem

Despite there being multiple algorithms to solve a problem, our aim should be to find the most efficient way (fewer requirements in terms of time and space) to do the same:

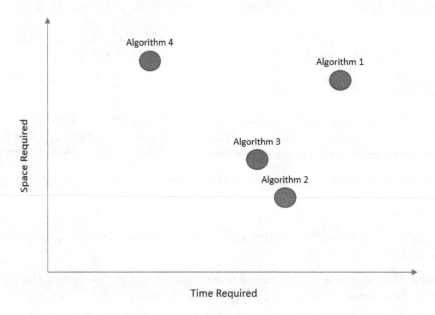

Figure 7.2 – Time and space complexity comparison for different algorithms

Oftentimes, there are multiple ways to solve a problem. However, we need to find the most efficient way to achieve this. To do this, we need to be able to quantify the performance of the different algorithms used and choose the best one. There are two things that are crucial while comparing the performance of algorithms, namely:

- **Time required**: This quantifies the amount of time required to run an algorithm to its completion as a function of length of the input.

  Time requirements can be defined as a numerical function $F(n)$, where $F(n)$ is measured as the number of steps, provided all steps consume the same amount of time.

  Let's say that the addition of two bits takes c seconds, hence, if we try to add n-bit integers, it will take $F(n) = n * c$ seconds. Hence, we can conclude that $F(n)$ has a linear growth as the size of the input increases.

- **Space required**: This quantifies the amount of memory space required by an algorithm in its life cycle. This required memory space has two components, namely:

- **Fixed part**: The space required for storing data and variables that are independent of the size of the problem we are trying to solve. This would include the predefined variables, constants, program size, and suchlike.

- **Variable part**: The space required for storing variables that are dependent on the size of the problem; for example, dynamic memory allocation, and recursion stack space.

Both time and space complexity are a function of the length of the input. In simpler terms, if the input size is larger, the algorithm will take a longer duration and require more memory space to run as compared to if the input is smaller. It is important to keep in mind that other factors, such as the hardware, processors, and operating system, play a crucial role in determining time and space complexity. However, for our purposes, we will only consider the *execution time* of an algorithm for analyzing it.

Let's try to understand the execution time with the help of an example.

For this example, the user will input a number and the algorithm will try to compare the input with a pre-existing list and give out an output accordingly:

Step 1: Ask the user to input a number.

Step 2: Compare the number with the numbers in the pre-existing list.

Step 3: If the input number matches any number in the list, then output Yes or else No. Remember that the algorithm looks for the match chronologically, in other words, the input number will be compared with the first number of the list, then the second, and so on.

Step 4: Display the time taken for the algorithm to run:

```
# a is a list containing some numbers
#We will compare the number input by user with the numbers in
  # this list

import timeit
tic=timeit.default_timer()

a=[1,2,3,4,5,6,7,8]
INPUT = input("Please input a number of your choice: ")
number = int(INPUT)

for i in range(len(a)):
    if a[i] == number:
        print("Yes", end=' ')
    else:
        print("No", end=' ')
print()

toc=timeit.default_timer()
time_elapsed = toc - tic
print("The time elapsed for this computation is: " + str(time_
  elapsed) + "seconds")
```

Output:

```
Please input a number of your choice: 1
Yes No No No No No No No
The time elapsed for this computation is: 2.3035541 seconds

Process finished with exit code 0
```

You can run the code and input a number of your choice and check for the runtime of the algorithm. For this algorithm, we had to import a Python library called timeit for measuring the time required for the algorithm to run. Documentation regarding the timeit library can be found here: https://docs.python.org/3/library/timeit.html.

An important thing to keep in mind is that the user input is being converted into an integer before comparing it with the integers in the array. If this is not done, then the algorithm would be comparing a string (input number) with integers (list) and hence would output only Nos.

Understanding Big-O Notation

Next, let's learn about **Big-O Notation**. Learning about this notation is crucial since it is used to describe the performance/complexity of an algorithm. This notation can be used to establish the relationship between the input to the algorithm and the steps required to execute the algorithm. **Notation**: O (relationship between the input and steps taken by the algorithm – denoted by "n").

For example: If there is a linear relationship between the input and the steps taken by the algorithm, then the Big-O notation will be *O(n)*. Similarly, for a constant relationship, the notation will be *O(constant)*.

The most frequently used Big-O notations are as follows:

Relationship between input and steps taken by algorithm	Big-O Notation
Constant	O(constant)
Linear	O(n)
Quadratic	$O(n^2)$
Cubic	$O(n^3)$
Exponential	$O(2^n)$
Logarithmic	O(log(n))
Log Linear	O(n*log(n))

Figure 7.3 – Big-O notation for different types of algorithms

We will now look into some of the complexities noted in the preceding table:

- **Constant complexity O(constant)**:

    The complexity of an algorithm is said to be constant if the steps required to execute the algorithm is constant despite the size of the input.

    Let's understand constant complexity with the help of an example in Python. We will make the algorithm do the following things:

    Step 1: Input a list.

    Step 2: Calculate the cube of the second item on the list:

    ```
    #Constant complexity function

    def constant_complexity(list):
        output = list[1]* list[1]* list[1]
    ```

```
        print("The end result after running the algorithm is:
           " + str(output))

constant_complexity([1,2,3,4,5,6,7])
```

Output:
```
The end result after running the algorithm is: 8
```

From the preceding example, we see that despite the length of the list that was input into the algorithm, it only does one thing – calculates the cube of the second item on the list (index numbers start from 0 in Python). Hence, the order of the algorithm is *O(2)* since the algorithm concludes in two steps despite the size of the input.

- **Linear complexity O(n)**:

The complexity of an algorithm is said to be linear if the steps required to execute the algorithm grow linearly with the size of the input.

Let's understand constant complexity with the help of an example in Python. We will make the algorithm do the following things:

Step 1: Input a list.

Step 2: Count the number of iterations taken to go through the entire list.

Step 3: Print the number of iterations:

```
#Linear complexity

def linear_complexity(list):
    for i in list:
        print("Iteration number " + str(i))

linear_complexity([1,2,3,4,5,6,7])
```

Output:
```
Iteration number 1
Iteration number 2
Iteration number 3
Iteration number 4
Iteration number 5
Iteration number 6
Iteration number 7

Process finished with exit code 0
```

From the preceding example, we can see that the number of iterations goes up as the length of the input goes up since the algorithm goes through each of the numbers in the input list – this is a linear relationship. Hence, we can represent this as $O(n)$, where n represents the number of steps taken by the algorithm.

- **Quadratic complexity $O(n^2)$**:

The complexity of an algorithm is said to be quadratic if the steps required to execute the algorithm grow quadratically with the size of the input. We will make the algorithm do the following things:

Step 1: Input a list.

Step 2: Go through two `for` loops:

```
#Quadratic complexity

def quadratic_complexity(list):
    count = 0
    for i in list:
        for j in list:
            count += 1
            print(str(count) + "\t|First for loop
                iteration: " + str(i), '\t|',
                    "Second for loop iteration:" + str(j))

quadratic_complexity([1,2,3,4])
```

Output:

1	|First for loop iteration: 1	| Second for loop iteration: 1
2	|First for loop iteration: 1	| Second for loop iteration: 2
3	|First for loop iteration: 1	| Second for loop iteration: 3
4	|First for loop iteration: 1	| Second for loop iteration: 4
5	|First for loop iteration: 2	| Second for loop iteration: 1
6	|First for loop iteration: 2	| Second for loop iteration: 2
7	|First for loop iteration: 2	| Second for loop iteration: 3

```
8     |First for loop iteration: 2    | Second for loop
       iteration: 4
9     |First for loop iteration: 3    | Second for loop
       iteration: 1
10    |First for loop iteration: 3    | Second for loop
       iteration: 2
11    |First for loop iteration: 3    | Second for loop
       iteration: 3
12    |First for loop iteration: 3    | Second for loop
       iteration: 4
13    |First for loop iteration: 4    | Second for loop
       iteration: 1
14    |First for loop iteration: 4    | Second for loop
       iteration: 2
15    |First for loop iteration: 4    | Second for loop
       iteration: 3
16    |First for loop iteration: 4    | Second for loop
       iteration: 4
```

When the `for` loop starts, the control of the program is first on the outer `for` loop. The first index in the list is 1. After this, the control of the program moves into the inner loop for execution. Here, each of the values in the list (`[1,2,3,4]`) is iterated over one at a time until the end of the list. `j` holds the value of 1 in the first iteration of the inner `for` loop followed by the execution of `print`, which outputs the number of iterations – `First for loop iteration:1 | Second for loop iteration: 1`, and moves to the next value in the list (which is 2). The control of the program is moved back to the outer loop once the execution of the inner loop is complete. This process is repeated until both the inner and outer loops have been executed completely.

The total number of steps performed is $n * n$ (for this case, the number of steps is 16), where n is the number of items in the input list.

The number of iterations will go up as the length of the input goes up, but in a quadratic manner – this is a quadratic relationship. Hence, we can represent this as $O(n^2)$, where n represents the number of steps taken by the algorithm.

- **Complexity of complex functions**:

  Next, we will look at an algorithm that does multiple things and try to figure out its Big-O notation. We will make the algorithm do the following things:

  Step 1: Print "Hello World!" six times à $O(6)$, since six steps are taken by the algorithm for this part.

Step 2: Use a `for` loop to go through the elements of a list and print them out à *O(n)*, since this is linear complexity and the number of steps taken is dependent on the number of elements in the list.

Step 3: Use a second `for` loop to go through the elements of a list and print them out a *O(n)*, since this is linear complexity and the number of steps taken is dependent on the number of elements in the list

The overall complexity of the algorithm is *O(6) + O(n) + O(n) = O(2n) + O(6)*:

```
#Complex function complexity

def complex_func (list):
    count = 0
    for i in range(6):
        count += 1
        print("Step: " + str(count) +   " \t Hello
           World!")

    for j in list:
        count += 1
        print("Step: " + str(count) +   " \t " + str(j))

    for k in list:
        count += 1
        print("Step: " + str(count) +   " \t " + str(k))

complex_func([1,2,3,4])
```

Output:
```
Step: 1     Hello World!
Step: 2     Hello World!
Step: 3     Hello World!
Step: 4     Hello World!
Step: 5     Hello World!
Step: 6     Hello World!
Step: 7     1
Step: 8     2
Step: 9     3
Step: 10    4
Step: 11    1
Step: 12    2
Step: 13    3
Step: 14    4
```

As you can see, the algorithm took 14 steps to complete the 3 tasks we wanted it to do. The complexity of the algorithm was found to be $O(2n) + O(6)$. However, when the size of the input list grows and becomes extremely large, then the constant become insignificant. This is the case because twice or half of infinity is still infinity. We can ignore the constants and the order of the algorithm will be $O(n)$ when the input list is extremely large. In short, we drop the non-dominant terms and the coefficients.

When do constants matter?

As regards the preceding example, we arrived at the conclusion that the order of the algorithm is $O(n)$ and dropped the constant terms, ($O(6)$). This is applicable when the problem size gets sufficiently large; the constant term does not matter. However, this means that two algorithms can have the same Big-O time complexity, even though one is always faster than the other. For example, suppose algorithm 1 requires n^2 time, and algorithm 2 requires $5*n^2 + n$ time. For both algorithms, if we ignore the constant terms, the Big-O notation is $O(n)$, even though algorithm 1 is faster than algorithm 2. In this case, the constants and low-order terms **do matter** in terms of which algorithm is faster.

However, it is important to note that constants **do not matter** in terms of the question of how an algorithm "scales" – in other words, how the algorithm's execution time changes when the problem size doubles or triples. Although an algorithm that requires n^2 time will always be faster than an algorithm that requires $10*n^2$, for **both** algorithms, if the problem size doubles, the actual time will quadruple.

When two algorithms have different Big-O time complexity, the constants and low-order terms matter only when the problem size is small. For example, if large constants are involved, then the linear time algorithm will be faster than the quadratic time algorithm:

n	100 * n	n²/100
10^2	10^4	10^2
10^3	10^5	10^4
10^4	10^6	10^6
10^5	10^7	10^8
10^6	10^8	10^{10}
10^7	10^9	10^{12}

Figure 7.4 – Linear and quadratic complexity for different input sizes

The preceding table shows the value of $100*n$ (linear in n) and the value of $n^2/100$ (quadratic in n) for some value of n. For values of n less than 10^4, the quadratic time is smaller than the linear time complexity. However, as the value of n increases beyond 10^4, the time complexity of quadratic is greater than the linear time complexity.

Now that we know how to come up with Big-O notation for an algorithm, let's represent this notation in a graph to make it clearer:

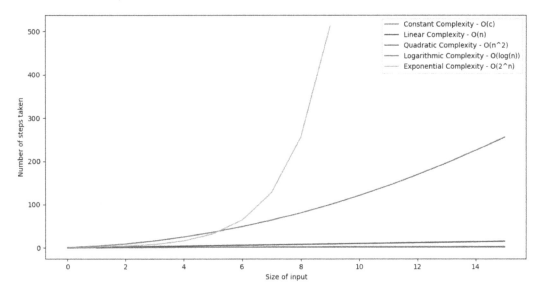

Figure 7.5 – Size of input versus the number of steps taken by an algorithm for different complexities

From the preceding graph, we can see how the number of steps taken by an algorithm for its execution is dependent on the size of the input for different kinds of time complexities.

In this section, we learned about Big-O notation and how to it calculate it for different complexities. In the next section, we will continue our discussion by learning about the complexity of algorithms with fundamental control structures.

Complexity of algorithms with fundamental control structures

In this section, we will learn about a crucial concept known as control structures. By the end of this chapter, you should have basic knowledge of control structures, their types, how they work, and their computational complexity.

Control structures are used to specify the direction of flow in programs. They are used to analyze and choose the direction in which the program flows, based on some parameters or conditions. In short, control structures are just the decision making that the computer makes. There are three basic types of fundamental control structures:

- Sequential flow
- Selection flow
- Repetitive flow

Let's understand each of these in turn.

Sequential flow

In this kind of flow, the algorithm flow depends on the series of instructions given to the computer, and the steps are executed in an obvious sequence. The sequence might be given by means of numbered steps explicitly. Also, it implicitly follows the order in which the steps are written. Most of the processing will generally follow this elementary flow pattern:

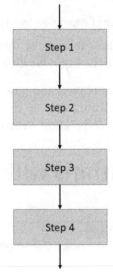

Figure 7.6 – Sequential flow

The complexity of this sequential flow is **constant**, since complexity is defined by the number of steps in an algorithm.

Selection flow

This type of flow involves several conditions or parameters that decide on one out of the several written steps. The structures that use these types of logic are known as conditional structures:

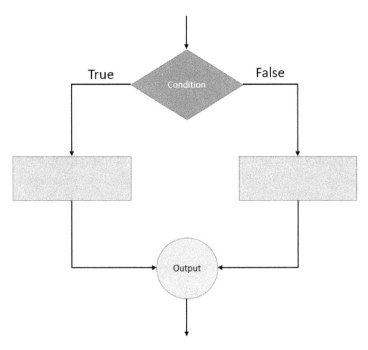

Figure 7.7 – Selection flow

A commonly used conditional in Python for selection flow is `if-elif-else`.

Let's recall some of the logical conditions used in mathematics that the Python programming language supports:

- **Equals**: a==b
- **Not Equals**: a!= b
- **Less than**: a < b
- **Less than or equal to**: a <= b
- **Greater than**: a > b
- **Greater than or equal to**: a >= b

Now that we have recalled some of the basic logical conditions, we can now apply these conditionals to better understand `if-elif-else` conditionals.

- **if-elif-else**:

    Decision making is required when we want to execute a certain section of the code if a certain condition is satisfied. The `if-elif-else` statement is used in Python for decision making.

    Let's see how it works with the help of an example. We will do the following for this example:

    Step 1: We will define two variables, `'a'` and `'b'`, and assign a value to each of them.

    Step 2: If a > b, then the algorithm will output something stating that a is greater than b.

    Step 3: If a < b, then the algorithm will output something else stating that a is less than b.

    Step 4: If a = b, then the algorithm will say that both numbers are equal:

    ```python
    #Complexity of if-elif-else statements

    a = 10
    b = 5
    if b > a:
        print("b is greater than a")
    elif b < a:
        print("b is less than a")
    else:
        print("a and b are equal")
    ```

    Output:

    ```
    b is less than a

    Process finished with exit code 0
    ```

    The Big-O notation for `if-elif-else` conditionals is $O(n)$ because in the worst case, the algorithm must go through all the n steps. If step 2 is true, then the algorithm will terminate after a single step, but if both steps 2 and 3 are false, then the algorithm will terminate after executing all the conditional statements.

Repetitive flow

This type of flow is used in the case of looping – where we are trying to run a piece of code a desired number of times or until a specified condition is applicable. There are two types of repetitive flow; let's discuss them in detail here:

- **Repeat-For Structure – For loop**:

    A for loop is used to iterate over a sequence that can be a list, tuple, dictionary, a set, string, and so on. With this loop, we can execute a set of statements, once for each item in a list, tuple, array, and suchlike:

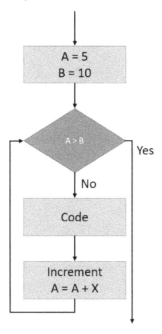

Figure 7.8 – Repetitive flow (for loop)

In the preceding diagram, two variables, A and B, are set to have a certain value. The for loop is run until A is less than B and the loop terminates once the condition $A > B$ is true. After going through the first iteration of the loop, the value of A is incremented by the number X.

Let's try to understand `for` loops better by going through an example in Python. For this example, we will perform the following steps:

Step 1: Print the list of fruit names :

```
#For loop

fruits = ["apple", "mango", "orange", "banana",
   "pomegranate"]
for x in fruits:
    print(x)
```

Output:

```
apple
mango
orange
banana
pomegranate
```

In the `for` loop, we are using x to represent the positions of the fruit names in the list fruits. For example, when x = 0, we are referring to `fruits[0]`, which represents the location of `apple`. Similarly, `fruits[2]` = "orange". It is important to remember that indices in Python start from 0. Here, we start the loop with x = 0, it prints the first element in the fruits list, then x is incremented to 1, where the second element in the list is printed out, and so on. The complexity of this `for` loop is $O(n)$. The loop executes n times (n being equal to the number in the list), so the sequence of statements also executes n times. Since we assume that the statements are of the order $O(1)$, the total time for the loop is $n * O(1) = O(n)$.

Now that we have an idea regarding a `for` loop, let's move on to learn about nested `for` loops and their complexity. So, what are nested loops? A nested loop is a loop inside a loop. The inner loop will be executed one time for each iteration of the outer loop'

In the previous example, we just printed out the names of the fruits. Now, let's add some adjectives to these fruit names by making use of nested `for` loops.

When the `for` loop starts, the control of the program is first on the outer `for` loop. The first adjective in the adjectives list, in this case `tasty`, is set into the value of y.

After this, the control of the program moves into the inner loop for execution. Here, each of the values in the `fruits` list is iterated over one at a time until the end of the list. x holds the value of `apple` in the first iteration of the inner `for` loop followed by the execution of `print(y, x)`, which outputs `tasty apple` and moves to the next value in the `fruits` list. The control of the program is moved back to the outer loop once the execution of the inner loop is complete, where the next value in the adjectives list is set as the value of y and the inner loop executes again, leading to the values to be printed as displayed in the output:

```
#Nested for loop

fruits = ["apple", "mango", "orange", "banana",
   "pomegranate"]
adjectives = ["tasty", "juicy", "fresh"]

foryin adjectives:
    for x in fruits:
        print(y, x)
```

Output:

```
tasty apple
tasty mango
tasty orange
tasty banana
tasty pomegranate
juicy apple
juicy mango
juicy orange
juicy banana
juicy pomegranate
fresh apple
fresh mango
fresh orange
fresh banana
fresh pomegranate
```

In the case of nested `for` loops, the complexity of the first loop is $O(n)$ and that of the second loop is $O(m)$. Since we do not know which one is bigger (for our example, we know that the inner loop was bigger), we can say that the complexity is $O(n+m)$. This can be written as $O(max(n,m))$.

- **Repeat-While Structure – While loop**:

With the `while` loop, we can execute a set of statements if a certain condition is `true`, until it stops being true:

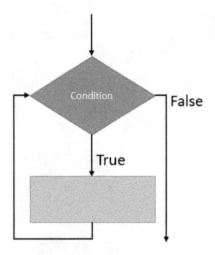

Figure 7.9 – Repeat-While structure (while loop)

Let's try to understand how to implement a `while` loop in Python. We will perform the following steps for this example:

Step 1: Set an index to a constant value.

Step 2: Print an index until a certain criterion is satisfied.

Step 3: Increment the index:

```
#While loop

i = 1
while i < 10:
    print("Step: " + str(i) + " The condition is
        satisfied")
    i += 1
```

Output:

```
Step: 1 The condition is satisfied
Step: 2 The condition is satisfied
Step: 3 The condition is satisfied
Step: 4 The condition is satisfied
Step: 5 The condition is satisfied
Step: 6 The condition is satisfied
Step: 7 The condition is satisfied
Step: 8 The condition is satisfied
Step: 9 The condition is satisfied
```

As regards the preceding example, we set the index $i = 1$ and use a `while` loop to print the statements telling us the step number and whether the condition $i < 10$ is satisfied. As you can see, the loop ended when the value of $i = 10$, and hence the algorithm stops at Step 10.

The complexity of the `while` loop is $O(n)$. This is because the complexity of a `while` loop depends on the loop control variable (which is *i* for this example) and how this variable is changing because the number of times the statements inside a loop get executed are dependent on this variable's behavior. For our example (above), the index `i` is being incremented linearly, in other words, the value of `i` increases by 1 for every step taken by the algorithm.

Now that we have learned about the complexity of common control structures and how the complexity is calculated, we will move on to study the complexity of common search algorithms in the next section.

Complexity of common search algorithms

Searching is a technique of selecting a certain portion of a dataset based on a certain set criterion. We use search algorithms in our day-to-day life when we search for something on the web that meets a certain word or phrase of our choice. Hence, to be able to search a data structure for required data is crucial in developing different kinds of applications.

In this section, we will discuss two search algorithms that are used in Python:

- Linear search algorithm
- Binary search algorithm

Linear search algorithm

This is the simplest kind of search algorithm to a sequential search problem. It simply checks the items in sequence until the desired item is found. This kind of search algorithm has already been illustrated using Example 2 (where we calculate the execution time of an algorithm). Let's look at a similar example to reinforce it, but this time we will write a function to carry out the linear search.

What is a Python function?

A function is a chunk of code that runs only when it is called. The user can pass inputs, known as parameters, into the function and then the function is executed to return a result.

For this example, we will be doing the following:

> **Step 1**: Write a function that contains a predefined set of lists of numbers.
>
> **Step 2**: Pass an input (number to compare to the list) into this function.
>
> **Step 3**: Print out the results.

This will return True if there is a match.

This will return False if there is no match:

```python
def linear_search(input):
    lists = [1, 2, 3, 4, 5, 6, 7, 8]
    number = int(input)

    for i in range(len(lists)):
        if lists[i] == number:
            print("True", end=' ')
        else:
            print("False", end=' ')
    print()

INPUT = input("Please input a number of your choice: ")
linear_search(INPUT)
```

Output:

```
Please input a number of your choice: 5
False False False False True False False False
```

Now that we know how a linear search algorithm works, let's consider the best-, worst-, and average-case scenarios:

- **Best Case**: This is the case that leads to the minimum number of steps executed. In the case of a linear search, the best-case scenario occurs when the target value (value that we are looking for) is present at the first index (0). The number of steps executed in this scenario is 1, hence the time complexity is a constant and the Big-O notation is $O(1)$.
- **Worst Case**: This is the case when the algorithm takes the maximum number of steps and, hence, the maximum amount of time. This scenario occurs when the algorithm is searching for an element that is present at the last index (n, let's say). The time complexity for this case will be $O(n)$ since the algorithm needs to take n steps (increasing linearly) before its termination.
- **Average Case**: The average case time can be found by dividing all the possible case timings (best and worse) by the number of cases. Hence, the average time complexity for a linear search is $O((n+1)/2)$.

Now that we have learned about best-, average-, and worst-case scenarios for linear search algorithms, let's learn about what these scenarios will be like for a binary search algorithm.

Binary search algorithm

This search algorithm requires a sorted sequence of a list and is based on the divide and conquer philosophy. It checks for the value in the middle of the list, repeatedly discarding the half of the list that contains values that are either all larger or all smaller than the desired value. If the midpoint contains the target, the algorithm immediately returns true. If this is not the case, then we determine if the target is less than the element at the midpoint or greater. If it is less, the high marker is adjusted to be one less than the midpoint, and if it is greater, we adjust the low marker to be one greater than the midpoint. In the next iteration of the loop, the only portion of the sequence that is considered is the elements between the low and high markers. This process is repeated until we find the target element, or the low marker becomes greater than the high marker. This is the termination condition and occurs when the target element is not found in the sorted sequence.

Let's look at an example to make this clearer. We will perform the following steps to come up with an algorithm:

list = [10, 20, 30, 40, 50, 60, 70, 80, 90, 100]

It is important to remember that the list needs to be sorted before a binary search is carried out:

Step 1: Compare the target value with the middle element of the list.

Step 2: If the middle element of the list matches the target element, then the index of the middle element is returned.

Step 3: If the target value is greater than the middle element in the list, then the target value can only lie in the right sub-list after the middle element.

Step 4: Otherwise (if the target value is less than the middle element), we look for a match in the sub-list that lies toward the left of the middle element:

```python
# Returns index of target (x) if present in the list
def binary_search(list, l, r, target_value):
    # Check base case
    if r >= l:

        mid_index = l + (r - l) // 2

        # If target element matches with the mid-element
          # of the list
        if list[mid_index] == target_value:
            return mid_index

        # If element is smaller than mid-element, then it
          # can only be present in left sublist
        elif list[mid_index] > target_value:
            return binary_search(list, l, mid_index - 1,
                target_value)

        # Else the element can only be present in right
          # sub-list
        else:
            return binary_search(list, mid_index + 1, r,
                target_value)

    else:
        # Element is not present in the array
        return -1


# Test list
```

```
list = [10, 20, 30, 40, 50, 60, 70, 80, 90, 100]
target_value = 100

# Function call
result = binary_search(list, 0, len(list) - 1, target_
    value)

if result != -1:
    print("Target element is present at index " +
        str(result) + " of the list")
else:
    print("Target element is not present in list")
```

Output:

```
Target element is present at index 9 of the list
```

Now that we know how a binary search algorithm works, let's consider the best-, worst-, and average-case scenarios:

- **Best Case**: This is the case that leads to the minimum number of steps executed. In the case of a binary search, the best-case scenario occurs when the target value (value that we are looking for) is present at the middle index of the list we are comparing with. The time complexity is a constant and the Big-O notation is $O(n)$.

- **Worst Case**: This is the case when the algorithm takes the maximum number of steps and, hence, the maximum amount of time. This scenario occurs when the algorithm is searching for an element that is *not present* in the list where we are looking to locate the target element. The time complexity for this case will be $O(\log n)$. This Big-O notation can be explained by the fact that the search keeps breaking the list into halves in each iteration. How many times do we have to divide a sub-list by 2 until we get our desired index? We can write this mathematically as follows:

$$\frac{n}{2^k} = 1$$
$$n = 2^k$$

Taking a logarithm on both sides, we get: $k = \log(n)/\log(2)$.

In the preceding computations, we have the following:

- **n** is the number of terms in the list.
- **k** is the number of times the sub-lists are divided into further smaller sub-lists.

- **Average Case**: The average case time can be found by dividing all the possible case timings (best and worse) by the number of cases. Hence, the average time complexity for a linear search is $O(\log n)$:

Type of search	Best Case	Worst Case	Average Case
Linear Search	O(1)	O(n)	O((n+1)/2)
Binary Search	O(1)	O(log n)	O(log n)

Figure 7.10 – Best, worst, and average case comparison for linear and binary searches

The advantage of binary searches is that not every item in the sequence must be examined before determining that the target is not in the list, which is the worst-case scenario. Since the sequence is sorted first, before proceeding with the binary search, each iteration of the loop can eliminate half of the values. In this way, the input size (the size of the list used for comparison) is repeatedly reduced by half during each iteration of the loop.

The binary search algorithm is more efficient as compared to a linear search since its worst-case time complexity is $O(\log n)$, which is better than $O(n)$ for a linear search.

In this section, we learned about the computational complexity of the two types of search algorithms used in Python, namely, linear and binary search algorithms. We also compared their best, worst, and average time complexity case scenarios.

Common classes of computational complexity

In this section, we will learn about some other common classes of computational complexity other than the constant, linear, quadratic, and suchlike complexities that have been discussed in the previous sections.

> *"Pretty well everybody outside the area of computer science thinks that if your program is running too slowly, what you need is a faster machine."*
>
> *– Rod Downey and Mike Fellows*

However, this is not the case, since some problems might require a brute-force search through a large class of cases that exponentially increases the number of steps required to solve the problem. An important distinction is often made between a **tractable** and **intractable** problem:

- **Tractable** problems make use of algorithms that take **polynomial time** (**P**) for their execution – time complexity is of the order $O(n^c)$, where c is any constant that belongs to the natural numbers.

Feasibly decidable kinds of problems are problems that can be solved by a conventional Turing machine in a number of steps that is proportional to a polynomial function of the size of the input.

- **Intractable** problems make use of algorithms that require exponential time for their execution – time complexity is of the order $O(2^n)$ or similar.

While this is the theoretical distinction, this might not always correspond to which problems can be solved faster in practice. For example, an exponential algorithm running in time $2^n/100$ might behave better than a polynomial algorithm running in time n^{1000}. The exponential functions might be faster for a very small number of steps; however, the polynomial time complexity will be faster when n is very large. Going back to *Figure 6.4*, we can see that the step size (and hence the time complexity) increases very rapidly for an exponential as compared to polynomial algorithms.

To put things into perspective, let's compare the run times of polynomial and exponential time complexities for different sizes of input and a step size of 10^{10} per second:

n	n^2 steps	2^n steps
2	0.00000002 msec	0.00000002 msec
5	0.00000015 msec	0.00000019 msec
10	0.00001 msec	0.0001 msec
20	0.0004 msec	0.10 msec
50	0.00025 msec	31.3 hours
100	0.001 msec	9.4 x 10^{11} years
1000	0.100 msec	7.9 x 10^{282} years

Figure 7.11 – Time taken by different algorithms for different step sizes

A third class of complexity class exists, called the **NP** type – *non-deterministic polynomial time*. This type consists of the problems that can be correctly decided by some computation of a non-deterministic Turing machine in a number of steps that is a polynomial function of the size of the input. These are the types of problems that are verifiable in polynomial time.

A famous conjecture states that **P** is properly contained in **NP** – in other words, *P Í NP*:

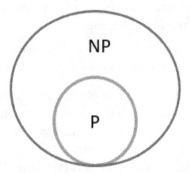

Figure 7.12 – P is properly contained in NP

Demonstrating the non-coincidence of these complexity classes remain important open problems in complexity theory.

In this section, we learned about some more complexity classes, including **P** and **NP**. We also looked at what tractable and intractable problems are and how exponential time complexity problems are computationally inefficient when the input size increases.

Summary

In this chapter, we learned about computer algorithms, and their complexities (time and space). We also discussed how these complexities vary based on the size of the input. We investigated the different types of time complexities, including constant, linear, quadratic, cubic, and exponential, along with their Big-O notations. We then looked into the complexities of fundamental control structures and discussed these with regard to three fundamental flow types – sequential, selection, and repetitive flow. The complexities of linear and binary search algorithms were discussed in addition to the best-, worst-, and average-case scenarios. Toward the end, we learned about some other kinds of time complexity types, such as P and NP.

With the knowledge acquired in this chapter, you will be well equipped to choose the right kind of algorithm to solve a certain problem. In the next chapter, we will be looking into terminology and notation for trees, graphs, and networks, as well as directed graphs and networks.

References

- Computational Complexity Theory (Stanford Encyclopedia of Philosophy):

  `https://plato.stanford.edu/entries/computational-complexity/`, Accessed: 2020-05-17.

- Computational Complexity Tutorial, COMSOC 2017, Ronald de Hann, URL: `https://staff.science.uva.nl/u.endriss/teaching/comsoc/2017/slides/comsoc-complexity-tutorial-2017.pdf`, Accessed: May 17, 2020.

8
Storage and Feature Extraction of Graphs, Trees, and Networks

The structures we will learn about in this chapter all stem from the idea of a graph, which is a pair of sets of nodes (called **vertices**) and connections (called **edges**) linking nodes together. As we will see in this chapter and the following chapters graphs, and their variations are useful for modeling many real situations and solving practical problems in computer and data sciences.

The following topics will be covered in this chapter:

- Understanding the terminology and notation of graphs, trees, and networks
- An overview of some ways graph and network models are used in real problems
- Efficient storage of graphs of networks in Python
- Using Python to extract features of graphs or networks

By the end of the chapter, you should be able to differentiate between graphs, trees, networks, and directed versions of them, be familiar with common applications of these structures as models for practical problems, efficiently store these structures in computer memory with Python, and use linear algebra to determine certain features within these structures.

> **Important Note**
> Please navigate to the graphic bundle link to refer to the color images for this chapter.

Understanding graphs, trees, and networks

We will start by defining graphs mathematically, along with any other related definitions, before moving on to consider common ideas about trees, networks, and directed graphs.

Definition: graph

A **graph** G has two parts. First, $V = \{v_1, v_2, ..., v_k\}$ is a set of **vertices**, also known as nodes. Second, E is a set of **edges**, each of which connects some pairs of nodes. We represent a graph as $G = (V, E)$.

An edge is represented mathematically as a set made up of the two vertices it connects. If there is an edge connecting nodes v_i and v_j, we will call this edge $e_{ij} = \{v_i, v_j\}$ and we say it is incident to vertices v_i and v_j.

An example of a graph follows with vertices $V = \{v_1, v_2, v_3, v_4, v_5, v_6\}$ and edges $E = \{e_{12}, e_{13}, e_{15}, e_{23}, e_{24}, e_{26}, e_{34}, e_{35}, e_{45}\}$:

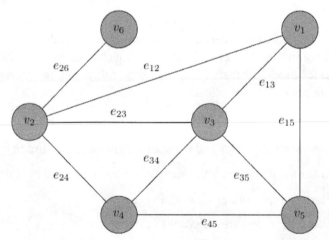

Figure 8.1 – A graph with six vertices, and nine edges connecting them

We can see, for example, the edge connecting vertex 3 (v_3) to vertex 4 (v_4) is called e_{34}. Note that, in general, it is common to leave out the edge labels from diagrams of graphs, but we can easily determine the name of each edge depending on which two vertices the edge connects.

Definition: degree of a vertex

A vertex v_i has **degree** n if it has exactly n edges incident to it. Mathematically, we write $d(v_i) = n$.

In other words, the degree of a vertex tells us how many edges are connected to the vertex. In the next example, we will count the degrees of each vertex in the graph from *Figure 8.1*.

Example: degrees of vertices

Consider the graph in *Figure 8.1*. We can easily find the degree of each vertex by counting the number of edges connected to it. We will find the following:

$$d(v_1) = 3, \quad d(v_2) = 4, \quad d(v_3) = 4$$
$$d(v_4) = 3, \quad d(v_5) = 3, \quad d(v_6) = 1$$

Notice the sum of the degrees of all the vertices is $3 + 4 + 4 + 3 + 3 + 1 = 18$, which happens to be two times the number of edges in the graph. It turns out this is true in general, as we'll prove next.

Theorem: sum of degrees

The sum of the degrees of all vertices in a graph $G = (V, E)$ equals twice the number of edges in G, $2|E|$. If the number of vertices is $|V| = n$, this means

$$\sum_{i=1}^{n} d(v_i) = d(v_1) + d(v_2) + \cdots + d(v_n) = 2|E|$$

Proof: An edge e_{ij} adds 1 to the degree of v_i and adds 1 to the degree of v_j. Therefore, each edge adds 2 to the sum of all degrees.

This fact is something we will use later in the chapter when we will store graphs and related models in Python to check that our data structure makes sense.

Next, we'll consider a special sort of graph where there is only one way to traverse are called paths, which we define next.

Definition: paths

A **path** is a graph $P = (V, E)$ where $V = \{v_1, ..., v_n\}$ and $E = \{e_{12}, e_{23}, ..., e_{n-1, n}\}$. The vertices v_1 and v_n are called endpoints of the path.

Here is an example of a path that is a subgraph of G from *Figure 8.1*, where $V = \{v_1, v_2, v_3, v_4, v_5\}$ and $E = \{e_{12}, e_{23}, e_{34}, e_{45}\}$:

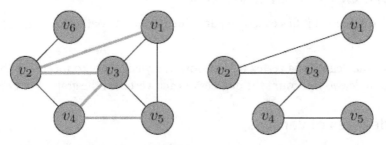

Figure 8.2 – On the left is the graph G; on the right is a path P taken from G

Next, we'll look at an idea closely related to paths – a graph where the starting vertex connects to the ending vertex, forming what is called a cycle.

Definition: cycles

If $P = (V, E)$ is a path, then a **cycle** is a graph with the same vertex set while the edge set is $E \cup \{e_{n1}\}$. In other words, it is a path with one additional edge connecting the endpoints.

The following diagram, *Figure 8.3*, shows a cycle that is a subgraph of the graph G from *Figure 8.1*:

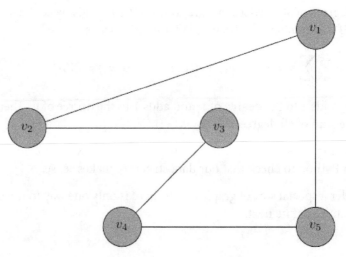

Figure 8.3 – A cycle that is a subgraph of G

As we mentioned before, a cycle is just like a path except the starting and ending vertices are connected. In this case, that means we added the edge e_{15} connecting vertex 1 to vertex 5 to the second graph from *Figure 8.2*.

With the idea of a cycle in hand, we can define trees, which are used anytime we want to create a hierarchy of objects such as operating systems, graphics, database systems, and computer networking.

Definition: trees or acyclic graphs

A **tree** or **acyclic graph** is a graph $G = (V, E)$ that has no cycles.

The graph in *Figure 8.4* is an example of a tree:

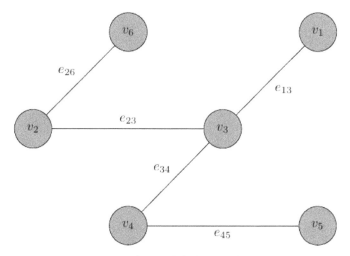

Figure 8.4 – An acyclic graph (otherwise known as a tree)

Notice there is no way to form a path from a vertex to itself without traversing the same edge more than once. This is what it means to not have a cycle.

Next, we define networks, which are like graphs, but the edges each have weights that may correspond to distances between cities in Google Maps, the cost of traveling from one vertex to another given fuel prices, or the weights of a deep learning structure like a neural network. The mathematical definition abstracts away those specifics so we can focus on the underlying ideas, which will then apply to many different problems.

Definition: networks

A **network** consists of three parts, $N = (V, E, W)$. As with graphs, V is the set of edges and E is the set of edges. In addition, each edge has a real-valued **weight**. Mathematically, we will write the set of weights as $W = \{w_{ij} \in \mathbb{R} : e_{ij} \in E\}$ and the weight of edge e_{ij} will be denoted by w_{ij}.

The following figure shows an example of a network with the same vertex and edge sets as the graph in *Figure 8.1* but with weighted edges:

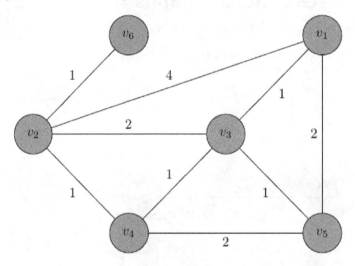

Figure 8.5 – A network with 6 vertices and 10 weighted edges

Just like the graph in *Figure 8.1*, this network has the vertex set $V = \{v_1, v_2, v_3, v_4, v_5, v_6\}$ and the edge set $E = \{e_{12}, e_{13}, e_{14}, e_{15}, e_{23}, e_{24}, e_{26}, e_{36}, e_{46}\}$, but it also has a set of weights given in the figure:

$W = \{w_{12}, w_{13}, w_{14}, w_{15}, w_{23}, w_{24}, w_{26}, w_{36}, w_{46}\} = \{2, 1, 4, 1, 1, 1, 1, 2, 2\}$.

The weights of a network may correspond to many different things in different applications, but some common examples are the distance between vertices, the capacity of the links to carry traffic between vertices, or the cost of making a connection between vertices. We will discuss these applications further in the next section.

Next, we'll continue on to directed versions of graphs where the edges do not simply connect vertices but have a specific direction from one vertex to another.

Definition: directed graphs

A **directed graph**, or **digraph**, $G = (V, E)$ is a set of vertices V and a set of directed edges E, which is a subset of the Cartesian product of V with itself,

$$E \subset V \times V = \{(v_i, v_j) : v_i, v_j \in V\}.$$

In contrast to (undirected) graphs, the edges here are **ordered pairs**, not just sets. The reason is that edges in this context have a direction. We call them **directed edges**.

In this context, $e_{12} = (v_1, v_2)$, an edge going from v_1 to v_2, but not the other direction. An edge going from v_2 to v_1 would be written $e_{21} = (v_2, v_1)$. In short, we have $e_{ij} \neq e_{ji}$ when we consider a directed graph. We will use arrows to the edges in diagrams of directed graphs. We can see an example in the following figure:

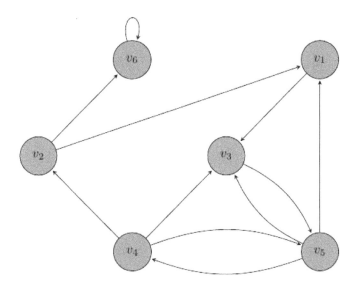

Figure 8.6 – A directed graph with 6 vertices and 11 directed edges

This digraph has the same six vertices as the previous graphs and its set of directed edges is $E = \{e_{13}, e_{21}, e_{26}, e_{35}, e_{42}, e_{43}, e_{45}, e_{51}, e_{53}, e_{54}, e_{66}\}$

These are directed edges, so in some cases, we have two edges—one in each direction—between two vertices; for example, e_{35} and e_{53}.

We discussed graphs and then generalized that idea by allowing edges to have directions. Now, we have talked about networks, which are like graphs but with weights. In a similar way, we can also allow the weighted edges of networks to become directed, creating what are called directed networks.

Definition: directed networks

A **directed network** $N = (V, E, W)$ is a network with directed edges.

Example: directed network

In this case, $|W| = |E|$ and W contains a weight for each directed edge, so the weight of edge e_{35} may be different than the weight of edge e_{53}. For example, the edge going from vertex 3 to vertex 5 has weight $w_{35} = 2$, but the weight of the edge going from vertex 5 to vertex 3 is $w_{53} = 1$, so the weights need not be the same in each direction, as we can see in *Figure 8.7*:

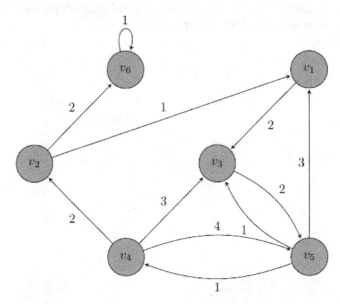

Figure 8.7 – A directed network with 6 vertices and 11 directed, weighted edges

In the preceding figure, we have a directed network where each directed edge has a weight. The vertex set and edge set are the same as in *Figure 8.6*, but here, we also have the set of weights,

$W = \{w_{13}, w_{21}, w_{26}, w_{35}, w_{42}, w_{43}, w_{45}, w_{51}, w_{53}, w_{54}, w_{66}\} = \{2, 1, 2, 2, 2, 3, 4, 3, 1, 1\}$

We will consider directed graphs and networks in some of the upcoming problems, but we will speak in the context of graphs and networks, not their directed variants unless it is otherwise specified.

Now that we have defined what graphs, trees, networks, and their directed variants are, we'll now consider a bit of terminology associated with these graph-based models in the next two definitions.

Definition: adjacent vertices

In a graph $G = (V, E)$, two vertices are called **adjacent** if an edge connects them. In other words, v_i and v_j are connected if $e_{ij} \in E$.

For example, in *Figure 8.1*, vertex v_3 is adjacent to v_1, v_2, v_4, and v_5 because there are edges attaching v_3 to each of those four vertices. However, it is *not* adjacent to v_6 since there is no edge connecting v_3 to v_6.

Lastly, we'll consider the idea of connected graphs and connected components of graphs.

Definition: connected graphs and connected components

Let $G = (V, E)$ be a graph. If G contains a path between every pair of vertices in V, then G is called a **connected graph**.

All the preceding figures are connected graphs because you can traverse some sequence of edges of the graphs to travel from any one vertex to any other vertex. This does *not* mean any two vertices are adjacent. Indeed, vertices v_4 and v_1 in *Figure 8.1* are not adjacent to one another, but some paths exist in G between them—for example, from v_4 to v_5 to v_1.

If a subgraph $G' = (V', E')$ of the graph G is connected and none of its vertices are connected to vertices outside G, it is called a **connected component** of the graph G:

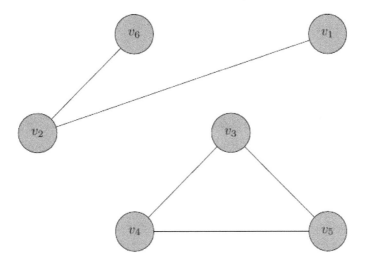

Figure 8.8 – A non-connected graph G with two connected components

In the preceding figure, there are two subgraphs of G we would like to consider:

1. $G1 = (\{v_1, v_2, v_6\}, \{e_{12}, e_{26}\})$
2. $G2 = (\{v_3, v_4, v_5\}, \{e_{34}, e_{35}, e_{45}\})$

If we choose a vertex from *G1*, there are no paths in *G* between this vertex and a vertex in the remainder of *G* (that is, in *G2*). The same would be true if we were to start with a vertex from *G2* and try to find a path to a vertex in *G1*. This means *G1* and *G2* are connected components.

We can note *G1* happens to be a tree, but not *G2* since it contains a cycle.

These ideas of connectedness and connected components also apply to networks. For directed graphs of networks, these notions also exist, but a path in each direction—from v_i to v_j and from v_j to v_i—must exist between each pair of vertices for the directed graph or network to be called connected.

In this section, we have seen many new terms and structures—the basis for all of them is the graph, which is simply a set of vertices and edges connecting some of the vertices together. Then, we saw that trees are graphs that include no cycles and networks are graphs where each edge has a weight. Next, networks were shown to be like graphs but with the addition of numerical weights for each edge. Lastly, we looked at the ideas of directed graphs or networks where the edges have a specific direction. Rather than an edge simply attaching vertex v_2 to vertex v_3, we have directed edges that go from v_2 to v_3 or go from v_3 to v_2 in directed graphs and networks. Throughout, we also encountered the ideas of degrees, paths, cycles, and connected graphs.

With all these new ideas in mind, let's take a look at some of the applications of these ideas in the real world!

Using graphs, trees, and networks

Graphs and the other similar structures we introduced in the previous section are versatile modeling tools. This section will be an overview of some of the most common areas where these structures are used in discrete mathematics. Note that some of these topics will be explored much more deeply in some forthcoming chapters.

In *Chapter 9*, *Searching Data Structures and Finding Shortest Paths*, we will learn how to search graphs (especially trees) to find certain features or characteristics. One application of these searches is in scheduling problems. For example, consider a directed graph where each vertex represents a task that needs to be done to complete a large project where a directed edge between task **A** and task **B** means task **A** must be completed before task **B**. In other words, the directed edge represents a dependency.

Clearly, there should be no cycles since that would lead to an infinite loop of tasks to complete! This means the directed graph would be a directed tree. There is a whole area of study of directed trees, which are also commonly called **directed acyclic graphs** (**DAGs**).

Searching such a directed tree can allow us to sort the tasks into orderings that allow the whole project to be completed efficiently. For example, consider the following figure, which shows a directed graph with the tasks involved in washing a car and their dependencies:

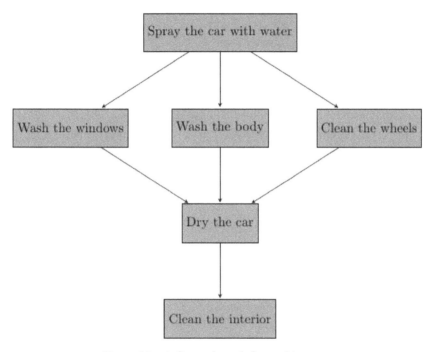

Figure 8.9 – A directed graph for washing a car

This figure sorts the tasks into a very easily readable structure that clearly shows the steps in the project. In general, we may have many tasks, each with a list of dependent tasks that must occur first. If they are not so neatly sorted or if a project is especially complex, scheduling the tasks can seem like a nearly impossible task. Searching these graphs allows such ordering, which is of tremendous use in project management.

Chapter 9, Searching Data Structures and Finding Shortest Paths, will look at the problems of finding minimum-weight paths between vertices in a network under various constraints. This is helpful for finding minimum-distance driving directions between two locations if weights represent distances between vertices, which may be cities or road intersections. In the following figure, we see the shortest path from v_1 to v_2 is highlighted in orange:

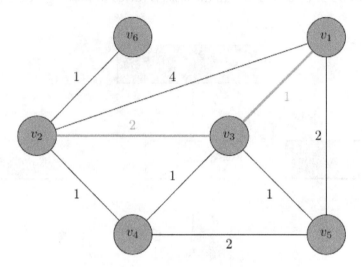

Figure 8.10 – The shortest path from v_1 to v_2

Note that there are many paths from v_1 to v_2:

- $P_1 = (\{v_1, v_2\}, \{e_{12}\})$, distance = $w_{12} = 4$
- $P_2 = (\{v_1, v_2, v_3\}, \{e_{13}, e_{23}\})$, distance = $w_{13} + w_{23} = 3$
- $P_3 = (\{v_1, v_2, v_3, v_4, v_5\}, \{e_{13}, e_{24}, e_{35}, e_{45}\})$, distance = $w_{13} + w_{24} + w_{35} + w_{45} = 5$

These are just a few of the paths from v_1 to v_2, but notice P_2, the one highlighted orange in *Figure 8.10*, is the shortest path with a total distance of *3* units.

Another problem related to pathfinding is routing traffic through a computer network—note the difference in the mathematical definition of a network given above and a network of computers. Networks in the mathematical sense sometimes model computer networks, but they can also model other things such as maps with roads connecting cities or an electrical grid connecting to all the customers in a geographical region.

One application of this problem is to find the cheapest way for an ISP to lay fiber optic cables to connect all the neighborhoods of customers they want to serve. We can see an example of a **minimum spanning tree** (**MST**) in the following figure:

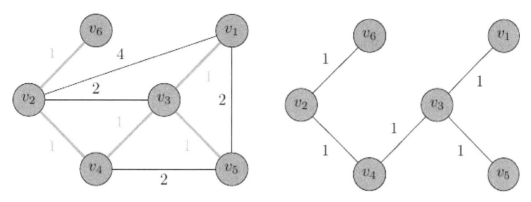

Figure 8.11 – On the left is N with the minimum spanning tree highlighted. The MST itself is on the right

We see that the MST is a connected subnetwork $N' = (V, E', W')$ with the same vertex set as the full network N but with a minimum sum of weights. The sum of weights on the left is *15*, but the sum of weights on the right is only *5*.

As we have seen in this section, graphs, trees, and networks can model many types of problems in scheduling, routing problems, and minimum spanning trees. This is only a small sampling of the common uses of these mathematical structures.

We will dive more fully into these applications in the next three chapters, but we cannot simply look at pictures for graphs modeling real-life problems. They tend to be quite large with hundreds or thousands of vertices and edges. The complexity of large graphs quickly exceeds our ability to analyze them mentally. So, before we can accomplish useful analysis, we need to learn how to store graphs, trees, and networks in NumPy arrays, which we will learn about next.

Storage of graphs and networks

In this section, we'll learn about a few ways graph structures are commonly stored in computer memory and their benefits and drawbacks, including adjacency lists, adjacency matrices, and weight matrices.

Definition: adjacency list

For a graph $G = (V, E)$, an **adjacency list** is an enumeration of the edges in a graph. In computer memory, we would store it as a list of pairs of vertex numbers.

Definition: adjacency matrix

For a graph $G = (V, E)$, an **adjacency matrix** for a graph is a binary matrix $A = (a_{ij})$. If $e_{ij} \in E$, then the number in row i and column j is $a_{ij} = 1$. Otherwise, it is 0.

In other words, the value in the ith row and j^{th} column of the adjacency matrix A, a_{ij}, is 1 if vertices v_i and v_j are adjacent. Otherwise, it is 0.

Example: an adjacency list and an adjacency matrix

For the graph G in *Figure 8.1*, we previously listed the edges as $E = \{e_{12}, e_{13}, e_{15}, e_{23}, e_{24}, e_{26}, e_{34}, e_{35}, e_{45}\}$. The adjacency list will simply be a list of each of these edges by the vertices they connect:

$$L1 = [[1, 2], [1, 3], [1, 5], [2, 3], [2, 4], [2, 6], [3, 4], [3, 5], [4, 5]]$$

While this is quite compact, it actually contains enough to describe the whole graph. The only risk is that a vertex with degree 0 will not be represented in the adjacency list. In most applications, this is unimportant. Vertices with no edges attached are not typically very interesting, but a separate list of vertices can be stored separately if it is important to the problem you are trying to solve.

Probably more common are adjacency matrices. The following matrix is the adjacency matrix for the graph in *Figure 8.1*. Note that $v_1, \ldots, v_6$ are not actually part of the matrix but are placed here as labels:

$$A_1 = \begin{array}{c} \\ v_1 \\ v_2 \\ v_3 \\ v_4 \\ v_5 \\ v_6 \end{array} \begin{array}{c} \begin{array}{cccccc} v_1 & v_2 & v_3 & v_4 & v_5 & v_6 \end{array} \\ \left[\begin{array}{cccccc} 0 & 1 & 1 & 0 & 1 & 0 \\ 1 & 0 & 1 & 1 & 0 & 1 \\ 1 & 1 & 0 & 1 & 1 & 0 \\ 0 & 1 & 1 & 0 & 1 & 0 \\ 1 & 0 & 1 & 1 & 0 & 0 \\ 0 & 1 & 0 & 0 & 0 & 0 \end{array} \right] \end{array}$$

Figure 8.12 – An adjacency matrix for the graph in Figure 8.1

Note, for example, that the number in the 4th row, 5th column is 1, which means v_4 is adjacent to v_5. To fill in the rest of the matrix, we use this same sort of logic to place a 1 in row i and column j if vertex v_i is connected to vertex v_j in the graph. All the rest of the numbers in the matrix will be zeros.

We can notice a few features of the adjacency matrix:

- The third row *[1 1 0 1 1 0]* means that v_3 is adjacent to v_1, v_2, v_4, and v_5 as we see in the preceding graph. The third column *[1 1 0 1 1 0]T* is the transpose of the third row because it represents the same thing – that v_3 is adjacent to v_1, v_2, v_4, and v_5.
- The transpose relationship is clearly true for the ith row and i^{th} column for any i.
- The diagonal is filled with zeros since no vertex is connected to itself. Self-connections are called loops, but our graph does not have any, so we can say G has no loops.
- The sum of the fifth row $1 + 0 + 1 + 1 + 0 + 0 = 3$ indicates the degree of v_5 is 3, $d(v_5) = 3$. In general, the sum of any row or column represents the degree of the corresponding vertex.

Notice each row and column have this property, so the transpose of A_1 is the same as A_1. Mathematically, we would write $A_{1T} = A_1$. In other words, $a_{ij} = a_{ji}$ for each i and j. Such a matrix is called a **symmetric matrix**. All adjacency matrices for non-directed graphs are symmetric.

Example: adjacency matrix for a non-connected graph

We can similarly write an adjacency matrix for the graph G in *Figure 8.8* as follows:

$$A_2 = \begin{bmatrix} 0 & 1 & 0 & 0 & 0 & 0 \\ 1 & 0 & 0 & 0 & 0 & 1 \\ 0 & 0 & 0 & 1 & 1 & 0 \\ 0 & 0 & 1 & 0 & 1 & 0 \\ 0 & 0 & 1 & 1 & 0 & 0 \\ 0 & 1 & 0 & 0 & 0 & 0 \end{bmatrix}$$

Figure 8.13 – An adjacency matrix for the graph in Figure 8.8

Once again, we have a symmetric matrix, which we showed must occur in an adjacency matrix. Here, we can quickly find the degree of each vertex by simply finding the row sums:

$$d(v_1) = 1, \quad d(v_2) = 2, \quad d(v_3) = 2$$
$$d(v_4) = 2, \quad d(v_5) = 2, \quad d(v_6) = 1$$

There is not a way to determine from observation that adjacency matrix A_2 corresponds to a graph that is not connected, but we will see in the next section that there is a way to use A_2 to determine this.

Definition: adjacency matrix for a directed graph

For a directed graph $G = (V, E)$, an adjacency matrix $A = (a_{ij})$ is a binary matrix where $a_{ij} = 1$ if there is a directed edge from vertex v_i to vertex v_j—that is, if $e_{ij} \in E$. All other elements of A are zeros.

Since $e_{ij} \in E$ does not mean $e_{ji} \in E$ for directed graphs, there is no reason to assume $a_{ij} = a_{ji}$ as in adjacency matrices of (undirected) graphs, so adjacency matrices for directed graphs are not symmetric in general.

> **Important note**
>
> In some books, authors define adjacency matrices for directed graphs differently. They let $a_{ij} = 1$ if there is an edge from v_j to v_i rather than the convention we have used above.

In the next example, we will find an adjacency matrix for a directed graph.

Example: adjacency matrix for a directed graph

The following matrix is the adjacency matrix for the directed graph G in *Figure 8.6*:

$$A_3 = \begin{bmatrix} 0 & 0 & 1 & 0 & 0 & 0 \\ 1 & 0 & 0 & 0 & 0 & 1 \\ 0 & 0 & 0 & 0 & 1 & 0 \\ 0 & 1 & 1 & 0 & 1 & 0 \\ 1 & 0 & 1 & 1 & 0 & 0 \\ 0 & 0 & 0 & 0 & 0 & 1 \end{bmatrix}$$

Figure 8.14 – An adjacency matrix for the graph in Figure 8.6

We can notice a few features of the adjacency matrix:

- The third row *[1 1 0 1 1 0]* means there are directed edges from v_3 to each v_1, v_2, v_4, and v_5 as we see in the preceding graph. The sum of the row is the number of edges leaving from v_3.
- The third column *[1 0 0 1 1 0]* means there are directed edges from each v_1, v_4, and v_5 to v_3. The sum of the column is the number of edges coming into v_3.
- $a_{66} = 1$ is on the diagonal since there is a loop going from v_6 to itself.

Note that, in contrast to the adjacency matrix we saw previously for an undirected graph, the adjacency matrix of this directed graph is *not* symmetric since we have directed edges.

We will come back to this example in the next section and use it to find some features of the directed graph corresponding to the adjacency matrix.

Example: storing an adjacency matrix in Python

To store an adjacency matrix in Python, it is smart to use a NumPy array as we saw in *Chapter 6, Computational Algorithms in Linear Algebra*. In the following code, we will store the adjacency matrices for the graphs in *Figure 8.1* and *Figure 8.8* as well as the directed graph in *Figure 8.6*:

```
import numpy

# Create an adjacency matrix for the graph in Figure 8.1
A1 = numpy.array([[0, 1, 1, 0, 1, 0], [1, 0, 1, 1, 0, 1],
                  [1, 1, 0, 1, 1, 0], [0, 1, 1, 0, 1, 0],
                  [1, 0, 1, 1, 0, 0], [0, 1, 0, 0, 0, 0]])

# Create an adjacency matrix for the graph in Figure 8.8
A2 = numpy.array([[0, 1, 0, 0, 0, 0], [1, 0, 0, 0, 0, 1],
                  [0, 0, 0, 1, 1, 0], [0, 0, 1, 0, 1, 0],
                  [0, 0, 1, 1, 0, 0], [0, 1, 0, 0, 0, 0]])

# Create an adjacency matrix for the directed graph in Figure
    # 8.6
A3 = numpy.array([[0, 0, 1, 0, 0, 0], [1, 0, 0, 0, 0, 1],
                  [0, 0, 0, 0, 1, 0], [0, 1, 1, 0, 1, 0],
                  [1, 0, 1, 1, 0, 0], [0, 0, 0, 0, 0, 1]])

# print the adjacency matrices
print("A1 =", A1)
print("\n A2 =", A2)
print("\n A3 =", A3)
```

This code outputs the matrices A_1, A_2, and A_3 that we wrote mathematically previously:

```
A1 = [[0 1 1 0 1 0]
 [1 0 1 1 0 1]
 [1 1 0 1 1 0]
 [0 1 1 0 1 0]
 [1 0 1 1 0 0]
```

```
    [0 1 0 0 0 0]]

A2 = [[0 1 0 0 0 0]
    [1 0 0 0 0 1]
    [0 0 0 1 1 0]
    [0 0 1 0 1 0]
    [0 0 1 1 0 0]
    [0 1 0 0 0 0]]

A3 = [[0 0 1 0 0 0]
    [1 0 0 0 0 1]
    [0 0 0 0 1 0]
    [0 1 1 0 1 0]
    [1 0 1 1 0 0]
    [0 0 0 0 0 1]]
```

As we see, the outputs are the exact adjacency matrices we found in the preceding examples, but they are now stored in computer memory in these NumPy arrays.

We will revisit these examples in the next section and use it to find some features of the graphs from *Figure 8.1*, *Figure 8.8*, and *Figure 8.6* corresponding with the adjacency matrices *A*, *B*, and *C*.

Efficient storage of adjacency data

An adjacency matrix is a little redundant and can, therefore, take up more memory than necessary. There are a few ways developers deal with this inefficiency.

Since adjacency matrices are always symmetric, sometimes code that uses adjacency matrices stores only the main diagonal of the matrix $(a_{11}, a_{22}, ..., a_{nn})$ and elements below the diagonal $(a_{ij}$ where $j \leq i)$. If we need an element that should be stored above the diagonal, say a_{24}, we can use symmetry to know it is equal to a_{42}, which is below the diagonal. In this way, we do not actually need to store a_{24} or any other element above the diagonal when the inefficiency is significant. When there are no loops in the graph, the diagonal of zeros also does not need to be stored.

These issues are unimportant if the amount of memory used is small, but for very large graphs, such as the web pages (vertices) and their link structure (edges) from a large website such as Reddit, the storage required can be very large, so cutting the storage space in half can be significant.

Further, if a graph has far more vertices than edges $|V| >> |E|$, an adjacency matrix will largely be filled with zeros. To be specific, the matrix is of size $|V|^2$ and there are $2|E|$ ones in the matrix, meaning there would be $|V|^2 - 2|E|$ zeros, which is a big number if $|V| >> |E|$. Storing all these zeros uses a lot of memory, often for not much benefit. In this situation, adjacency lists are sometimes preferred, as we need to store only $2|E|$ values, the two endpoints of each edge.

Next, let's look at the corresponding ideas for networks.

Definition: weight matrix of a network

For a network $N = (V, E, W)$, a cost matrix is a matrix $W = (w_{ij})$ – that is, where the number in row i and column j is the weight w_{ij} of the edge connecting vertices v_i and v_j if the edge exists. If there is no edge between v_i and v_j, we set $w_{ij} = 0$ in the matrix.

Since it may be the case that the weight of the edge from vertex v_i to vertex v_j is not equal to the weight of the edge from vertex v_j to vertex v_i in directed networks (or maybe the second edge does not even exist!), there is no reason to assume $w_{ij} = w_{ji}$ as in weight matrices of (undirected) networks, so weight matrices for directed networks are not symmetric in general.

> **Important note**
>
> In some sources, weight matrices are referred to by various other names depending on what the networks are being used to model—distance matrices, cost matrices, or even simply adjacency matrices. We will use a weight matrix exclusively.

We will consider an example of the weight matrix of a specific network in the next example.

Example: weight matrix of a network

The weight matrix from the network shown in *Figure 8.5* is given here:

$$\mathbf{W}_1 = \begin{bmatrix} 0 & 4 & 1 & 0 & 2 & 0 \\ 4 & 0 & 2 & 1 & 0 & 1 \\ 1 & 2 & 0 & 1 & 1 & 0 \\ 0 & 1 & 1 & 0 & 2 & 0 \\ 2 & 0 & 1 & 2 & 0 & 0 \\ 0 & 1 & 0 & 0 & 0 & 0 \end{bmatrix}$$

Figure 8.15 – A weight matrix for the network in Figure 8.5

We constructed the matrix by noting, for example, the weight of the edge connecting v_1 and v_2 is 4, so $w_{12} = w_{21} = 4$, and continuing in the same way to fill in the remainder of the numbers.

Weight matrices of (undirected) networks share some properties with adjacency matrices of are symmetric and zeros occur in positions of the matrix corresponding to any two vertices that are not connected by an edge. That is, vertices that are not adjacent.

Definition: weight matrix of a directed network

For a directed network $N = (V, E, W)$, a cost matrix is a matrix $W = (w_{ij})$ – that is, where the number in row i and column j is the weight w_{ij} of the edge going from vertex v_i to vertex v_j if the edge exists. If there is no edge going from v_i to v_j, we set $w_{ij} = 0$ in the matrix.

Let's consider an example from a directed network we saw in an earlier section to make this idea clearer.

Example: weight matrix of a directed network

Let's find the weight matrix for the network shown in *Figure 8.7*:

$$W_2 = \begin{bmatrix} 0 & 0 & 2 & 0 & 0 & 0 \\ 1 & 0 & 0 & 0 & 0 & 2 \\ 0 & 0 & 0 & 0 & 2 & 0 \\ 0 & 2 & 3 & 0 & 4 & 0 \\ 3 & 0 & 1 & 1 & 0 & 0 \\ 0 & 0 & 0 & 0 & 0 & 1 \end{bmatrix}$$

Figure 8.16 – A weight matrix for the network in Figure 8.7

We constructed the matrix by noting, for example, the weight of the edge going from v_1 to v_3 is 3, so $w_{13} = 3$ and continuing in the same way to fill in the remainder of the numbers. Unlike the undirected network cost matrix, this one is not symmetric since, for example, $w_{31} = 0 \neq w_{13}$ since there is no edge going from v_3 to v_1.

Example: storing weight matrices in Python

To store a weight matrix in Python, it is smart to use a NumPy array as we saw in *Chapter 6, Computational Algorithms in Linear Algebra*. In the following code, we will store the weight matrices for the network in *Figure 8.5* and the directed network in *Figure 8.7*:

```
import numpy
```

```
# Create a weight matrix for the network in Figure 8.5
W1 = numpy.array([[0, 4, 1, 0, 2, 0], [4, 0, 2, 1, 0, 1],
                  [1, 2, 0, 1, 1, 0], [0, 1, 1, 0, 2, 0],
                  [2, 0, 1, 2, 0, 0], [0, 1, 0, 0, 0, 0]])

# Create a weight matrix for the directed network in Figure 8.7
W2 = numpy.array([[0, 0, 2, 0, 0, 0], [1, 0, 0, 0, 0, 2],
                  [0, 0, 0, 0, 2, 0], [0, 2, 3, 0, 4, 0],
                  [3, 0, 1, 1, 0, 0], [0, 0, 0, 0, 0, 1]])

# Print the weight matrices
print("W1 =", W1)
print("\n W2 =", W2)
```

And the code has output:

```
W1 = [[0 4 1 0 2 0]
 [4 0 2 1 0 1]
 [1 2 0 1 1 0]
 [0 1 1 0 2 0]
 [2 0 1 2 0 0]
 [0 1 0 0 0 0]]

 W2 = [[0 0 2 0 0 0]
 [1 0 0 0 0 2]
 [0 0 0 0 2 0]
 [0 2 3 0 4 0]
 [3 0 1 1 0 0]
 [0 0 0 0 0 1]]
```

As we see, the weight matrices have been stored as NumPy arrays in computer memory by Python. It is then prepared for analysis with Python.

Now that we have learned how to store graphs as adjacency matrices and networks as weight matrices, we are prepared to look at some approaches to extract features from graphs and networks in Python.

Feature extraction of graphs

In this section, we will learn how to find features of graphs from their adjacency matrices using some methods from linear algebra we learned in *Chapter 6, Computational Algorithms in Linear Algebra*—especially matrix sums and matrix multiplication.

We will learn how to find the degrees of vertices, count the paths between vertices of a specified length, and find the shortest paths between vertices of graphs.

Degrees of vertices in a graph

In this subsection, we will learn how to find the degrees of vertices with Python. As we mentioned in the previous section, the row (or column) sums of an adjacency matrix give the degrees of each vertex.

We do these calculations in Python:

```
# Find the degrees of each vertex of the graph in Figure 8.1

# Using column sums
print(numpy.sum(A1, axis=0))

# Using row sums
print(numpy.sum(A1, axis=1))
```

Note that we use the `sum()` function from NumPy where the first input is the adjacency matrix A_1 of the graph in *Figure 8.1* and the second is the axis, which specifies whether it should sum the rows or sum the columns. In the first one, we use `axis=0`, so it computes the column sums. In the second, we use `axis=1`, so it computes the row sums. The output follows:

```
[3 4 4 3 3 1]
[3 4 4 3 3 1]
```

Of course, the two agree with one another for an undirected graph since their adjacency matrices must be symmetric. And, these numbers agree with the degrees we found by inspection. Of course, this counting by inspection for every vertex on a real, large graph would be infeasible to do manually.

For a directed graph, we must realize the adjacency matrix is constructed differently. A_1 in row i, column j means there is a directed edge going from vertex v_i to vertex v_j. So, if we add all the numbers in row i, this will give the number of edges leaving from v_i, sometimes called the out-degree of v_i. Practically, we can compute this out-degree for each vertex by computing row sums.

In contrast, to find the number of edges entering v_j, sometimes called the in-degree of v_j, we need to add up column j. In general, we need to compute column sums to get the in-degrees of the vertices.

We will implement both for adjacency matrix A_3 corresponding to the directed graph in *Figure 8.6* in Python here:

```
# Find out-degrees for each vertex in the directed graph in
  # Figure 8.6
outdegrees = numpy.sum(A3, axis=1)
print(outdegrees)

# Find in-degrees for each vertex in the directed graph in
  # Figure 8.6
indegrees = numpy.sum(A3, axis=0)
print(indegrees)

print(numpy.sum(outdegrees))
print(numpy.sum(indegrees))
```

This code gives the out-degrees and then the in-degrees for the vertices:

```
[1 2 1 3 3 1]
[2 1 3 1 2 2]
11
11
```

In this case, we computed the row and column sums just like we did for the preceding undirected graph, but here, there is a different interpretation when we are considering directed graphs. The in-degree and out-degree of each vertex differ in general since a vertex may have different amounts of edges entering it and exiting from it. However, the sum of the in-degrees, 11, equals the sum of the out-degrees since each exiting edge must enter some vertex. This number is precisely the number of edges in the directed graph.

The next few ideas we will see have to do with counting the number of paths between vertices.

The number of paths between vertices of a specified length

Consider the adjacency matrix for the graph in *Figure 8.1*, A_1. Each element of the matrix, a_{ij}, is 1 if there is an edge connecting vertex v_i to vertex v_j and 0 otherwise. In other words, an element of the matrix is 1 if there is a path of length 1 between the 2 vertices and 0 otherwise.

It turns out, multiplying adjacency matrices by themselves reveals some features of graphs that may not be so easy to determine by inspection, especially for large graphs. For example, suppose we multiply the adjacency matrix by itself:

$$\mathbf{A}_1^2 = \mathbf{A}_1 \mathbf{A}_1$$

The number in row i, column j comes from computing the dot product between row i of the first A_1 by column j of the second A_1. For example, if $i = 2$ and $j = 3$, we have

$a_{21}a_{13} + a_{22}a_{23} + a_{23}a_{33} + a_{24}a_{43} + a_{25}a_{53} + a_{26}a_{63} = (1)(1) + (0)(1) + (1)(0) + (1)(1) + (0)(1) + (1)(0) = 2$

Since these are binary values, if $a_{2j}a_{j3} = 1$, then there are both edges between v_2 and v_j and an edge between v_j and v_3, meaning there is a path from v_2 to v_j to v_3. Otherwise, at least one of these edges is not in the graph so the path would not exist.

The sum of these for all j, as we computed earlier, is, therefore, the number of paths with two edges between v_2 and v_3. Each other element of $A_1 A_1$ is constructed in the same way, so each element of the product gives the number of two-edge paths between each pair of nodes as follows:

$$\mathbf{A}_1^2 = \begin{bmatrix} 3 & 1 & 2 & 3 & 1 & 1 \\ 1 & 4 & 2 & 1 & 3 & 0 \\ 2 & 2 & 4 & 2 & 2 & 1 \\ 3 & 1 & 2 & 3 & 1 & 1 \\ 1 & 3 & 2 & 1 & 3 & 0 \\ 1 & 0 & 1 & 1 & 0 & 1 \end{bmatrix}$$

Figure 8.17 – The adjacency for the graph in Figure 8.1 multiplied by itself

There are several details of the squared adjacency matrix that correspond to some features of the graph:

- The matrix is symmetric since, for example, the number of paths from v_2 to v_3 is the same as the number of paths from v_3 to v_2.

- The diagonal elements equal the degree of the vertices. For example, the number in row 3, column 3 is 4 since each edge traversed twice makes up a two-edge path from v_3 to itself.

It turns out the pattern of counting paths continues for higher powers of the adjacency matrix. When we multiply by A_1 for a third time, we will get the number of three-edge paths between each pair of vertices. In general, we have the following theorem.

Theorem: powers of adjacency matrices

For a graph $G = (V, E)$ with adjacency matrix A, the number in row i, column j in the matrix A_n is the number of paths with n edges between vertex v_i and vertex v_j in the graph.

Matrix powers in Python

We learned how to multiply matrices with the Python function numpy.dot() in *Chapter 6, Computational Algorithms in Linear Algebra*, which we could use multiple times with a loop to repeatedly multiply by a matrix, but here, we will learn a better way is to use the numpy.linalg.matrix_power function, also from NumPy.

For example, let's recreate the calculation of the number of two-edge paths between each pair of vertices in the graph depicted in *Figure 8.1* with Python and also find the number of three-edge paths between each pair of vertices by taking the third power of the adjacency matrix A_1:

```
# Find the second power of adjacency matrix A1
print(numpy.linalg.matrix_power(A1,2))

# Find the third power of adjacency matrix A1
print("\n", numpy.linalg.matrix_power(A1,3))
```

Then, the output is as follows:

```
[[3 1 2 3 1 1]
 [1 4 2 1 3 0]
 [2 2 4 2 2 1]
 [3 1 2 3 1 1]
 [1 3 2 1 3 0]
 [1 0 1 1 0 1]]

[[4 9 8 4 8 1]
 [9 4 9 9 4 4]
 [8 9 8 8 8 2]
 [4 9 8 4 8 1]
 [8 4 8 8 4 3]
 [1 4 2 1 3 0]]
```

The code finds the second and third powers of the adjacency matrix and prints them.

In general, we tend to see larger numbers in the third power because there are more three-edge paths between most pairs of vertices than there are two-edge paths.

One notable exception is that the element in row 6, column 6 is 0 because there are no three-edge paths from v_6 to itself because any path must start with the edge e_{62} and end with the edge e_{26}, so adding any one additional edge cannot form a path returning to v_6 since v_2 has no self-connection.

We can use this idea to determine the shortest path between two vertices, as we will do next.

Theorem: minimum-edge paths between v_i and v_j

The minimum n such that the number in row i, column j in the matrix A_n is positive is the number of edges in the minimum-edge path from v_i to v_j. In other words, n is the shortest distance from v_i to v_j.

This theorem is self-evident from the previous theorem. Since the number in row i, column j of $A_1, A_2, A_3, \ldots$ represents the number of paths from v_i to v_j with 1 edge, with 2 edges, with 3 edges, and so on. Therefore, the first power where the number is not 0 is the shortest path that exists between those vertices.

Example: paths between nodes in Figure 8.8

Consider the graph in *Figure 8.8*. Let's find the number of paths of different lengths between some pairs of nodes with Python. We will write a loop to calculate powers of the matrix from $n = 1$ to $n = 6$ and print the number of paths of each length between some given vertices. See the following code for these operations:

```
# Print the number of paths from v1 to v6 of each length from 1
    # to 6
for counter in range(1,7):
    A2counter = numpy.linalg.matrix_power(A2,counter)
    print("There are", A2counter[0,5], "paths of length",
        counter, "from v1 to v6")

# Print the number of paths from v2 to v3 of each length from 1
    # to 6
for counter in range(1,7):
    A2counter = numpy.linalg.matrix_power(A2,counter)
    print("There are", A2counter[1,2], "paths of length",
        counter, "from v2 to v3")
```

And the output is the following:

```
There are 0 paths of length 1 from v1 to v6
There are 1 paths of length 2 from v1 to v6
```

```
There are 0 paths of length 3 from v1 to v6
There are 2 paths of length 4 from v1 to v6
There are 0 paths of length 5 from v1 to v6
There are 4 paths of length 6 from v1 to v6

There are 0 paths of length 1 from v2 to v3
There are 0 paths of length 2 from v2 to v3
There are 0 paths of length 3 from v2 to v3
There are 0 paths of length 4 from v2 to v3
There are 0 paths of length 5 from v2 to v3
There are 0 paths of length 6 from v2 to v3
```

From the first loop, we see the shortest path from v_1 to v_6 is 2 and it seems only odd-length paths exist between the two, corresponding to the number of times we will traverse the two-edge path v_1-v_2-v_6. In contrast, there are no paths of length 6 or less between v_2 and v_3. This suggests what we can see by inspection—there are no paths between v_2 and v_3.

This brings the section to an end, now that we have learned how to extract degrees, the number of paths between nodes, and short paths on graphs with Python.

Summary

In this chapter, we began by introducing the ideas of graphs, directed graphs, networks, and directed networks along with some common language used to describe them. Next, we introduced a few ways in which these structures are used for modeling practical problems, many to be investigated more deeply in the forthcoming chapters.

After this, we moved on to consider ways in which graphs and networks can be stored in computer memory with Python. Especially popular are adjacency matrices and adjacency lists for graphs and weight matrices for networks. In the last section, we showed many features of graphs from adjacency matrices, such as degrees of vertices, the number of paths between pairs of vertices, and the length of the minimum-edge paths between the vertices.

Altogether, this chapter has defined graphs, trees, networks, and the directed types of these structures, established some common vocabulary on these topics, familiarized you with some practical applications of each, shown how they can be stored in computer memory—most frequently in the forms of adjacency or cost matrices, and how to extract some features of the graphs from these matrices using NumPy and Python code.

These new skills will serve you well and open doors to a very effective type of modeling for practical problems using graphs, trees, and networks. In *Searching Data Structures and Finding Shortest Paths*, we will focus on algorithms for traversing graphs and trees to detect more complex features of the graphs. These algorithms have many practical applications in web crawling for Google, finding driving directions on MapQuest, locating sources of files in peer-to-peer networks such as BitTorrent, and recommending friends to users on Facebook.

9
Searching Data Structures and Finding Shortest Paths

This chapter will discuss the searching techniques of graph, tree, and network data structures and practical applications of graph searches. We will introduce and analyze two popular algorithms for related problems: **depth-first search** (**DFS**) for graph searches and Dijkstra's algorithm for finding the shortest paths between vertices in networks. Both are introduced on small graphs to build intuitive understanding, and Python implementations are written that can scale up to real-world problems.

In this chapter, we will cover the following topics:

- Searching graph and tree data structures
- Depth-first search algorithm
- The shortest path problem and variations of the problem

- Finding shortest paths with brute force
- Dijkstra's algorithm for finding shortest paths
- Python implementation of Dijkstra's algorithm

By the end of this chapter, you will be able to explain the purpose of searching, implement the DFS methods, understand shortest path problems and their variants, and implement Dijkstra's algorithm to find shortest paths.

> **Important note**
> Please navigate to the graphic bundle link to find the color images for this chapter.

Searching Graph and Tree data structures

In the previous chapter, we learned about graphs and trees. As we progress through the chapter, keep in mind that whenever we refer to graphs, this includes trees because trees are simply graphs that have no cycles. The topic of this section is the idea of searching graphs. This simply means to travel along the edges of a graph to locate paths to destination vertices. This sounds like a simple thing to do, but we hope to do it as efficiently as we can because many real-world graphs are huge.

There are many reasons why we might want an algorithm to traverse the graph to find vertices. For example, suppose you want to send a message over the internet to five of your friends living in five different cities. There certainly will be no direct connection between your device and your friends' devices, so the message must follow multiple paths from vertex to vertex through networked devices until it reaches your friends. Networked devices connect and disconnect from each other from time to time, so it is not possible for us to store a permanent graph representing the network. This means the paths must be mapped out at the time you want to send the message. This is what a graph search can do.

Now, determining along which path to send the message is a different question. As the graph search maps the paths, we may want to choose the paths that take the least time to deliver the messages or paths that flow through connections that are not congested. We will learn about finding the shortest path later in the chapter, but for now, it suffices to say that a graph search is frequently an important part of solving such problems.

This is the norm. Graph searches do not do too much on their own, but they tend to be used as subroutines in complex algorithms that solve many problems, such as finding shortest paths and minimum spanning trees, detecting connected components of graphs, analyzing network flow, matching vertices from one group with another, or large scheduling problems where tasks have complex relationships.

Depth-first search (DFS)

In short, graph searches traverse a graph to map its structure. In this section, we will learn about an algorithm to accomplish such a search. Mapping out the structure of a graph can be important on its own, but it is a sub-problem that algorithms must solve in order to solve larger problems in graphs, as we have discussed. The DFS algorithm is quite possibly the most common approach for graph searches; it is an efficient method, and it is used as a subroutine in many more complex algorithms.

DFS starts at a source vertex, traverses the first available edge to visit another vertex, and repeats this until there are no edges leading to unvisited vertices—that is, until it has gone as deep as possible. At this time, it backtracks to the last vertex that has unvisited neighbors and takes another trip from that vertex through as many unvisited vertices until it reaches another dead end. It then backtracks and travels to unvisited vertices again and again until all the vertices connected to the source have been visited.

Let's pursue this method on the following small graph so that we can understand the idea well. We will start at v_1 and explore the graph using DFS.

Note that it was not specified how to choose a path, so we will arbitrarily move to the lowest-numbered vertex when we have more than one option. We will color vertices and edges within the current path orange and previously visited vertices and previously traversed edges will be green:

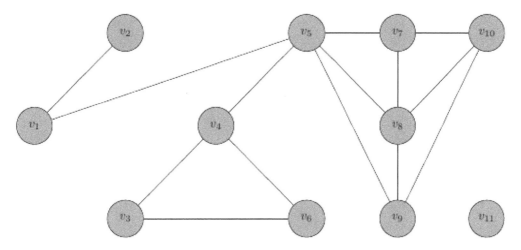

Figure 9.1 – A graph

Step 1: The first step will go to v_2, which is not adjacent to any vertices we have not yet visited, so it stops:

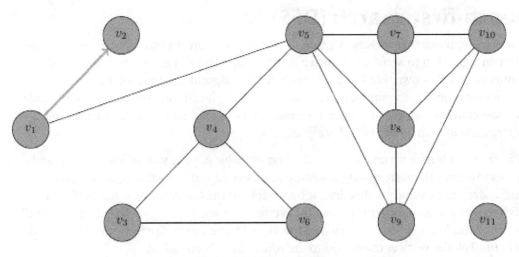

Figure 9.2 – Step 1 of DFS

Step 2: We backtrack to node v_1, and then follow paths until we reach a dead end once again. This will take us from v_1 to v_5 to v_4 to v_3 to v_6, which has no unvisited neighbors, so we stop:

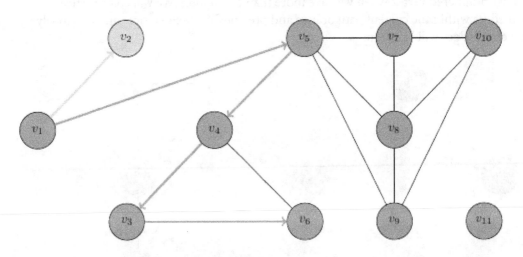

Figure 9.3 – Step 2 of DFS

Step 3: We backtrack all the way to v_5 because it is the latest one in the orange path with unvisited neighbors and take a path to v_7 to v_8 to v_9 to v_{10} and stop:

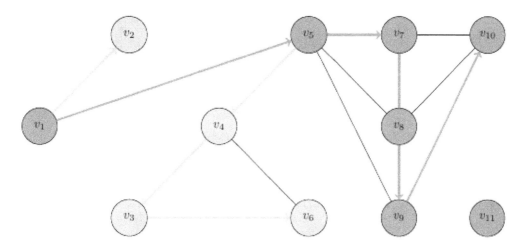

Figure 9.4 – Step 3 of DFS

Finally, all the vertices connected to source v_1 are colored in our diagram, indicating all have been visited, so the graph search is complete.

The list of vertices this DFS would produce is as follows:

$$v_1, v_2, v_5, v_4, v_3, v_6, v_7, v_8, v_9, v_{10}$$

Notice vertex v_{11} has not been visited because it is not connected to the source vertex. In general, DFS will not leave a connected component of the source vertex. For a graph with multiple connected components, you would have to run DFS once within each component if you wanted to visit all the vertices.

Now, let's move on to write an implementation of the DFS algorithm in Python.

A Python implementation of DFS

Of course, for large, practical problems, we cannot simply apply the algorithm by hand! Instead, let's write an implementation of the DFS algorithm in Python.

We will write a function called DFS that will take an input of an adjacency matrix of a graph and will return all the vertices connected by a path to the source vertex.

We will present it in pieces and explain as we go. First, we have some documentation listing what the function does and outlines its inputs and outputs:

```
# Depth First Search
#
# INPUTS
```

```
# A - an adjacency matrix. It should be square, symmetric, and
# binary
# source - the number of the source vertex
#
# OUTPUTS
# vertexList - an ordered list of vertices found in the search
```

Next, we define the functions with inputs of an adjacency matrix and source vertex, subtract the source by 1 since Python counts from 0, find the number of vertices in the graph, and initialize several data structures, including a binary array to store which vertices have been visited, a stack to be used in the algorithm, and a vertex list the algorithm will fill in:

```python
def DFS(A, source):
    # reduce the source by 1 to avoid off-by-1 errors
    source -= 1

    # find the number of vertices
    n = A.shape[0]

    # initialize the unvisited vertex set to be full
    unvisited = [1] * n

    # initialize a queue with the source vertex
    stack = [source]

    # initialize the vertex list
    vertexList = []
```

Then, take the last vertex in the stack and add it to the vertex list if it has not been visited, and add all unvisited neighboring vertices to the end of the queue. Repeat this until the stack is empty. And, lastly, return the vertex list:

```python
    # while the stack is not empty
    while stack:
        # remove the just-visited vertex from the stack and
          # store it
        v = stack.pop()

        # if v is unvisited, add it to our list and mark it as
          # visited
        if unvisited[v]:
            # save and print the number of the newly visited
              # vertex
            vertexList.append(v)

            # mark the vertex as visited
            unvisited[v] = 0

        # iterate through the vertices
        for u in range(n - 1, 0, -1):
            # add each unvisited neighbor to the stack
            if A[v,u] == 1 and unvisited[u] == 1:
                stack.append(u)

    return vertexList
```

Now that the code is written, let's test it on the example we did previously by hand just to confirm it works as intended. We will need to save the adjacency matrix first:

```
# Save the adjacency matrix for the graph in Figure 9.1
A = numpy.array([[0, 1, 0, 0, 1, 0, 0, 0, 0, 0, 0],
                 [1, 0, 0, 0, 0, 0, 0, 0, 0, 0, 0],
                 [0, 0, 0, 1, 0, 1, 0, 0, 0, 0, 0],
                 [0, 0, 1, 0, 1, 1, 0, 0, 0, 0, 0],
                 [1, 0, 0, 1, 0, 0, 1, 1, 1, 0, 0],
                 [0, 0, 1, 1, 0, 0, 0, 0, 0, 0, 0],
                 [0, 0, 0, 0, 1, 0, 0, 1, 0, 1, 0],
                 [0, 0, 0, 0, 1, 0, 1, 0, 1, 1, 0],
                 [0, 0, 0, 0, 1, 0, 0, 1, 0, 1, 0],
                 [0, 0, 0, 0, 0, 0, 1, 1, 1, 0, 0],
                 [0, 0, 0, 0, 0, 0, 0, 0, 0, 0, 0]])
```

Next, let's run the DFS algorithm with source vertex 1 just like we did by hand before. We will also add 1 to each of the numbers in the vertex list since we have counted from 1 unlike Python:

```
# Run DFS on the graph with adjacency matrix A and source 1
vertexList = DFS(A,1)

# Add 1 to the vertex numbers
[x + 1 for x in vertexList]
```

The output is as follows:

```
[1, 2, 5, 4, 3, 6, 7, 8, 9, 10]
```

When we applied the algorithm by hand, note that we found the exact same list in the exact same order. Clearly, the code is replicating what we were able to do by hand except it runs almost instantly, so our DFS implementation is a great success!

In this section, we have learned what the DFS algorithm is, discussed some of its applications, applied it by hand to an example, wrote a Python implementation of the algorithm, and showed that it matches the results of our example.

The remainder of the chapter focuses on a very practical problem: finding the shortest path between two vertices in a network or weighted graph.

The shortest path problem and variations of the problem

In this section, we shift our focus to a different graph-related problem: finding the shortest paths between vertices in a network. As we will discuss, this is a problem that is important for routing problems, such as finding the shortest route to travel in a car to a destination or finding the fastest way to deliver a message over a computer network. Shortest path problems have even been used to determine how to use the thrusters on small fleets of deep-space research satellites to move them into very precise positions in relation to one another with minimal fuel usage so that they could work in unison to capture images of stars.

For graphs with unweighted edges, we have previously solved this problem. Let's review this simpler problem and its solution briefly before continuing to the more general problem on networks (that is, weighted graphs). In *Chapter 8, Storage and Feature Extraction of Graphs, Trees, and Networks*, we found a way to find the minimum-edge path, or shortest path, between nodes v_i and v_j on a graph or directed graph. It was simply the smallest number, n, such that the nth power of the adjacency matrix, A_n, has a positive value in row i, column j. This was obvious since the number in this position gives the number of v_i-to-v_j paths of length n.

This result is useful because it allows us to find the shortest distance between nodes in graphs and directed graphs in an efficient way since matrix multiplication is computationally cheap, with computational complexity below $O(n^3)$.

Shortest paths on networks

However, a problem with much wider applicability is finding the shortest path between nodes in a network where the edge weights represent the distance between the nodes. This is a very important problem. It can allow map apps such as MapQuest, Google Maps, or Waze to find a route with the shortest-distance path between two cities, which is something many of us use every day! An equivalent problem is finding the shortest distance to supply electricity from a power source to a customer through nodes in an electrical grid. Given that there is more loss of energy over longer distances, a smart grid would keep these distances small to efficiently use the energy generated by power stations.

Beyond Shortest-Distance Paths

Beyond these examples, it is also possible to interpret weights as something other than distances. For example, if we use Google Maps to find a path to a certain address, we might be more interested in how long it takes to reach the address than the distance traveled. Traveling to the address on foot might yield the shortest distance, but if the distance is measured in tens or hundreds of miles, the shortest distance might be quite useless! Instead, we may assign weights to the network corresponding to the *time* it takes to travel between nodes. A related problem is used by driving apps such as Waze to use real-time traffic data to provide better estimates of the time it takes to drive between two locations. Here, again, the distance is not the most important factor in finding optimal driving directions—we would like to have the shortest-time route:

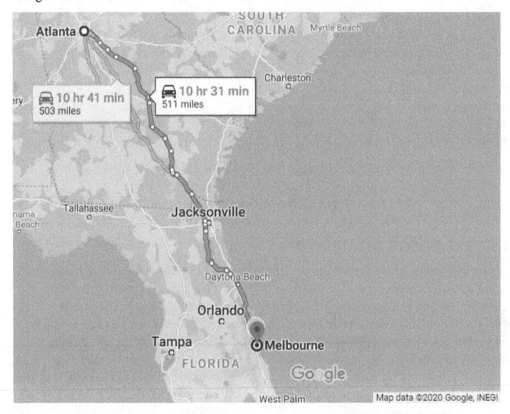

Figure 9.5 – In this map, we see two routes: one 503 miles and the other 511 miles

Notice, in *Figure 9.5*, how Google Maps recommends the longer-distance path instead of the shortest path. This is because its goal is to give the route with the shortest *time*, which happens to be slightly longer in distance. There are many reasons why this could be true: the shorter route may have more traffic, lower speed limits, or more traffic lights.

Treating weights as times opens a whole new set of applied problems where we are interested in shortest-time paths. For example, if you want to send a text message from your phone to another person's PC, we would like to choose a path over a network to send the message from your phone to your friend's PC. Here, we are far more interested in the latency, or lag-time, in delivering the message than the distance the signal must travel. Finding shortest-time paths allows us to have intelligent policies for sending traffic over computer networks or the internet.

Another option is to use weights that represent the cost of adding an edge to a path. For example, perhaps it costs money for a traveler to traverse an edge, such as costs for fuel or wages for a truck driver in the context of driving, and we may want to find the minimum-cost path, even if the distance and time are not minimal. Similarly, if we would like to build a road connecting two cities through some intermediate nodes, the costs of building each stretch of roadway between each pair of nodes will be different depending on not only the distance between the nodes but also the terrain between them, the distance of transporting materials and workers, and many other considerations.

Whether we seek shortest paths in the context of distance, time, cost, or some other consideration, we have seen that they all break down to the same problem in the context of networks—seeking the minimum sum of weights for a path connecting two nodes we select. This merging of so many different problems into one abstract problem in terms of networks displays the power of mathematics to generalize and solve many problems at once.

Shortest Path Problem Statement

We can see that there are reasons to let the weights represent very different measurements in different applied problems, so let's abstract away from specific assumptions on what they represent and formalize the problem statement for finding shortest paths on networks.

Let $N = (V, E, W)$ be a network, where $V = \{v_1, v_2, \ldots, v_n\}$ is the set of vertices, E is the set of edges connecting pairs of vertices, and W is the set of weights of the edges. We will seek the shortest path from vertex v_i to vertex v_j. That is, we want to find a set of edges connecting v_i to v_j with a minimal sum of edge weights.

Note that there may be many different paths between a given pair of nodes or there may be no paths between them. If there are paths between them, there may be multiple paths with a minimal sum of weights. As such, we should bear in mind that it is a problem where solutions may not exist, there may be a unique solution, or there may be multiple solutions.

For most practical purposes, only the possibility that there is no solution is especially important. What this means is that v_i is not connected to v_j by any path. Let's learn how to check that v_i is connected to v_j using a method from the previous chapter.

Checking whether Solutions Exist

Recall that we could find the path with minimal edges from v_i to v_j by exponentiating the adjacency matrix until the value in row i, column j is non-negative. Of course, this will never happen if v_i is not connected to v_j, but we will generally know how many edges, $|E|$, are in a network. Therefore, if A^n has a 0 in row i, column j for all cases of $n \leq |E|$, then there will be no path between these nodes and we will know that there is no shortest path because, even if we use all the edges in the graph, it does not contain a path from vertex v_i to v_j, so they must not be connected. In the case that they are connected, there exists a path and so there exists a minimal-weight path.

Thus, before using an algorithm to find the shortest path, it is a good idea to confirm whether any path exists first by exponentiating the adjacency matrix until we confirm. Let's write a Python function to do this check. We will simply check whether the vertices are adjacent and, if not, exponentiate the adjacency matrix one power at a time until we can confirm that a path exists. Or, if we reach $A^{|E|}$ without detecting a path, we know there are no paths from v_i to v_j. In this case, we will know that there is no solution to the shortest path problem, so we can avoid the trouble of searching for it!

Our function will return `True` if there is a path and print the length of the path. The function will return `False` and print a notice that no path was found:

```
import numpy

# create a function that returns True if vertex i and vertex j
  # are connected in the graph represented by the input
    # adjacency matrix A
def isConnected(A, i, j):
    # initialize the paths matrix to adjacency matrix A
    paths = A

    # find the number of vertices in the graph
    numberOfVertices = A.shape[0]

    # find the number of edges in the graph
    numberOfEdges = numpy.sum(A)/2

    # if vi and vj are adjacent, return True
```

```python
        if paths[i-1][j-1] > 0:
            print('Vertex', i, 'and vertex', j, 'are adjacent')
            return True

    else:
        # run the loop until we find a path
        for pathLength in range(2, numberOfVertices):
            # exponentiate the adjacency matrix
            paths = numpy.dot(paths, A)

            # if the element in row i, column j is more than 0,
            # we found a path
            if paths[i-1][j-1] > 0:
                print('There is a path with', pathLength,
                      'edges from vertex', i, 'to vertex', j)
                return True

        # found no paths, the vertices are not connected
        if pathLength == numberOfEdges:
            print('There are no paths from vertex', i, 'to',
                  'vertex', j)
            return False
```

Since we have written a function, there is no output from this code as it is written, but we can run it by inputting a specific adjacency matrix for a graph along with the vertex numbers *i* and *j*. Writing a function gives us the advantage of being able to reuse it as much as we like with different inputs to determine whether different vertices are connected.

To test our code, let's use it to find some path lengths between vertices on a small graph we can easily determine visually and check whether the code replicates these facts. Recall the following graphs from *Chapter 8, Storage and Feature Extraction of Graphs, Trees, and Networks*:

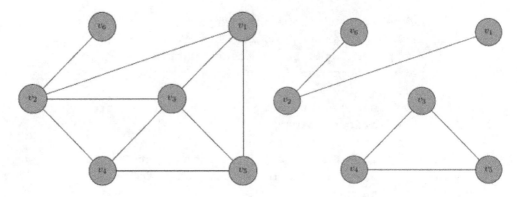

Figure 9.6 – Graph G1 (left) and graph G2 (right)

Let's call the graph on the left G_1 and the graph on the right G_2. Of course, it is okay if the graphs are actually networks with edge weights, but the weights are unimportant to determining whether or not two vertices are connected in the network:

```
# create an adjacency matrix for the graph G1
A1 = numpy.array([[0, 1, 1, 0, 1, 0], [1, 0, 1, 1, 0, 1],
                  [1, 1, 0, 1, 1, 0], [0, 1, 1, 0, 1, 0],
                  [1, 0, 1, 1, 0, 0], [0, 1, 0, 0, 0, 0]])

# check if various vertices are connected
print(isConnected(A1, 1, 4))
print(isConnected(A1, 2, 3))
print(isConnected(A1, 5, 6))
```

Here, we entered the adjacency matrix for graph G_1 and checked whether several pairs of vertices are connected. The output of the code is as follows:

```
There is a path with 2 edges from vertex 1 to vertex 4
True

Vertex 2 and vertex 3 are adjacent
True

There is a path with 3 edges from vertex 5 to vertex 6
True
```

Clearly, these outputs match the facts we can easily see from the graph: there is a two-edge path from v_1 to v_4, there is an edge connecting v_2 and v_3, and there is a three-edge path from v_5 to v_6.

With graph G2, the code should output False for some choices of vertices. Let's try it out:

```
# create an adjacency matrix for graph G2
A2 = numpy.array([[0, 1, 0, 0, 0, 0], [1, 0, 0, 0, 0, 1],
                  [0, 0, 0, 1, 1, 0], [0, 0, 1, 0, 1, 0],
                  [0, 0, 1, 1, 0, 0], [0, 1, 0, 0, 0, 0]])

print(isConnected(A2, 1, 6))
print(isConnected(A2, 2, 5))
print(isConnected(A2, 1, 4))
```

The output is as follows:

```
There is a path with 2 edges from vertex 1 to vertex 6
True

There are no paths from vertex 2 to vertex 5
False

There are no paths from vertex 1 to vertex 4
False
```

Again, the code replicates the facts we can easily see from looking at the diagram of graph G_2: vertices v_1 and v_6 are connected, vertices v_2 and v_5 are not connected, and vertices v_1 and v_4 are not connected.

Now that we have a method to verify solutions exist before we look for them, we will discuss a method to find the shortest path in a small problem.

> **Important note**
> Note that, for large networks, this check for connectedness is somewhat expensive to run. In this case, we would skip straight to searching for the shortest paths, although we must realize that the search will fail if vertices v_i and v_j are not connected.

Finding Shortest Paths with Brute Force

As we laid out in the previous section, we will seek a path from vertex v_i to vertex v_j with a minimal sum of edge weights. Let's look at the prospects of finding the shortest paths using brute force.

For example, consider the following network that we discussed in *Chapter 8, Storage and Feature Extraction of Graphs, Trees, and Networks*. We will let V be the set of vertices, E be the set of edges, and W be the set of weights:

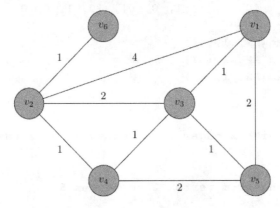

Figure 9.7 – A network

An example problem that we will try to solve is to find the shortest path from v_1 to v_2. There are many paths between these two vertices, which we list as follows along with their lengths:

Paths from v_1 to v_2	Path Lengths
$v_1 - v_2$	4
$v_1 - v_3 - v_2$	1 + 2 = 3
$v_1 - v_3 - v_4 - v_2$	1 + 1 + 1 = 3
$v_1 - v_3 - v_5 - v_4 - v_2$	1 + 1 + 2 + 1 = 5
$v_1 - v_5 - v_3 - v_2$	2 + 1 + 2 = 5
$v_1 - v_5 - v_4 - v_2$	2 + 2 + 1 = 5
$v_1 - v_5 - v_4 - v_3 - v_2$	2 + 2 + 1 + 2 = 7

Figure 9.8 – All the paths from v_1 to v_2 and their lengths, excluding paths that revisit the same vertex

From this full list of paths from v_1 to v_2, we can easily see that the shortest paths are the ones in the highlighted rows with lengths of 3 units, either taking a path from v_1 to v_3 to v_2 or a path from v_1 to v_3 to v_4 to v_2.

Notice that these short paths contain more edges than the minimal-edge path that simply goes directly from v_1 to v_2, which has a length of 4 units. Of course, the path length is not necessarily dependent on the distance of taking the path and we should not expect the shortest paths to necessarily have the fewest number of edges.

Here, we have simply listed all possible paths, but for a large graph, this could be incredibly expensive to do. For example, suppose a graph with n vertices is complete, meaning there is an edge between every pair of vertices. So, vertex v_1 has $n - 1$ incident edges. Vertex v_2 has $n - 2$ incident vertices, plus the edge from v_1 to v_2, which was already counted. Vertex v_3 has $n - 3$ incident edges, plus two edges from v_1 and v_2. Continuing this pattern, we eventually find just 1 uncounted edge incident to v_{n-1} and all the edges incident to vn have been counted. Then, the number of edges altogether is as follows:

$$1 + 2 + 3 + \cdots + n - 1$$

According to an inductive proof in *Chapter 2, Formal Logic and Constructing Mathematical Proofs*, the sum of the first $n - 1$ non-negative integer is as follows:

$$\frac{(n-1)n}{2}$$

If the graph has, for example, 100 vertices, then there would be *(100)(99)/2 = 4,950* edges, and so there could be millions of distinct paths from one vertex to another!

So, just how many paths does this mean there would be between a pair of edges? Suppose we want to count the number of paths from v_i to v_j that contain k additional vertices. There are $|V| - 2$ edges to choose from, so as we learned in *Chapter 4, Combinatorics Using SciPy*, the number of such paths is as follows:

$$\binom{|V| - 2}{k} = \frac{(|V| - 2)!}{k!\,(|V| - 2 - k)!}$$

This is the case for any k value between 0 and $|V| - 2$. Therefore, the number of paths in our 100-vertex complete graph containing 5 other vertices is as follows:

$$\binom{100 - 2}{5} = \binom{98}{5} = \frac{98!}{5!\,(98 - 5)!} = 67{,}910{,}864$$

But, of course, there's no reason why there would be five additional vertices. There could be *2, 3, 4, ..., 98* vertices, meaning the number of paths is as follows:

$$\binom{98}{0} + \binom{98}{1} + \binom{98}{2} + \cdots + \binom{98}{98}$$

It is a little beyond the scope of this book, but this sum is known to equal the following:

$$2^{98} \approx 3.17 \times 10^{29}$$

As a result, a brute-force approach such as this is clearly limited! It would take an entirely unrealistic amount of time to test this many paths. Also, this 100-vertex graph is quite small, especially when you consider the fact that one practical application—map apps such as Google Maps—place a vertex at every intersection between pairs of roads within entire cities and beyond. This would mean there are over 12,000 vertices in New York City alone!

While this brute-force method is easy to understand, it is clearly infeasible, so we need a more strategic approach to solve the problem in a useful amount of time. We need an efficient way to find the shortest paths between specified vertices on networks or directed networks, assuming, of course, that a solution exists. This is what Dijkstra's algorithm does, so let's learn about it!

Dijkstra's Algorithm for Finding Shortest Paths

In this section, we will learn about Dijkstra's algorithm for finding the shortest paths, consider the process in simple terms, and apply the algorithm by hand to a small network.

The most common algorithm for finding the shortest paths on a network is Dijkstra's algorithm. It was named after the Dutch computer scientist *Edsger W. Dijkstra*, who constructed it in 1956, but since computing was such a new field at the time, there were so few academic journals dedicated to computing that he did not publish his findings until 1959.

We will first learn about the method in intuitive terms using the small network from *Figure 9.5* so that we can understand the ideas behind the approach. This understanding is important because there are many variations of the algorithm and we hope you will learn to adapt it to solve your own problems!

Just like the previous section, we will seek the shortest path from v_1 to v_2. Since it is a small network, we were able to find that there are such paths using brute force. The paths were as follows:

$$v_1 - v_3 - v_2 \text{ and } v_1 - v_3 - v_4 - v_2$$

Each of these has a length of 3 units. We will actually construct the short paths from v_1 to every other vertex in the network along the way to find the shortest path from v_1 to v_2, as this is how Dijkstra's algorithm is typically implemented.

Dijkstra's algorithm

We will start at vertex v_1 and traverse the graph as we carry out Dijkstra's algorithm. Along the way, we will maintain two lists: vertices we have visited in the method and vertices we have not visited in the method. The visited set is empty initially and the unvisited set will have all the vertices in it, as we can see:

- Visited vertices = { }
- Unvisited vertices = $\{v_1, v_2, v_3, v_4, v_5, v_6\}$

In our problem, the starting point is node v_1, which we call the source. Dijkstra's algorithm follows the following pattern:

- **Initialization**: Set the distance to each vertex from the source to infinity and the distance to itself as 0.
- Visit the nearest unvisited adjacent vertex with the shortest known distance from the source (ties can be broken arbitrarily):

    a) If any distances through the current vertex from the source are shorter than the known distances, update the shortest distances.

    b) For any replaced shortest distances, record the "previous vertex" as the current vertex.

    c) Add the current vertex to the visited vertices list.
- Repeat the work from the previous bullet point until we have visited all of the vertices.

In the end, Dijkstra's algorithm will give the shortest paths from the source v_1 to *every* other vertex in the graph, which is much more than we asked for, but in many problems, we would like to know more than just the one path between specified vertices.

Note that Dijkstra's algorithm is called a **greedy algorithm** because it chooses the cheapest path from the source at each step. Of course, building on the shortest existing path need not lead to the best path, but sometimes it does. If we get lucky, we will find the shortest path with some of these early choices. If not, then the algorithm will still eventually find the solution, but it may have to backtrack a significant number of times before it finds the solution, which is not too fast, but it is still *far* faster than a brute-force algorithm could ever hope to be on a problem on a realistically large scale.

216   Searching Data Structures and Finding Shortest Paths

Applying Dijkstra's Algorithm to a Small Problem

Let's see whether we can follow these steps for the preceding small network! We will explain each step, draw an updated network highlighting the current vertex and the new edges to be investigated to be incorporated into a shortest path, update a table of shortest distances and previous vertices, and maintain the lists of visited and unvisited vertices.

Step 0 (initialization): Set the shortest path distance to each vertex to infinity, ∞, except we set the distance from the source to itself to be 0:

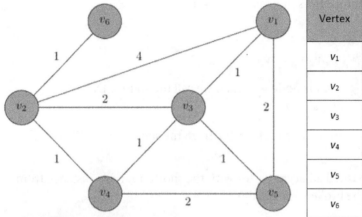

Vertex	Shortest Distance	Previous Vertex
v_1	∞	
v_2	∞	
v_3	∞	
v_4	∞	
v_5	∞	
v_6	∞	

Visited Vertices	Unvisited Vertices
	$v_1, v_2, v_3, v_4, v_5, v_6$

Figure 9.9 – Step 0 of Dijkstra's algorithm

Step 1: Add v_1 to the set of visited vertices. Find the distance from the source to all adjacent nodes in the unvisited vertices set. If the distance is shorter than the current distance, save it:

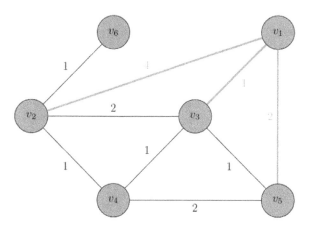

Vertex	Shortest Distance	Previous Vertex
v_1	0	
v_2	4	v_1
v_3	1	v_1
v_4	∞	
v_5	2	v_1
v_6	∞	

Visited Vertices	Unvisited Vertices
v_1	v_2, v_3, v_4, v_5, v_6

Figure 9.10 – Step 1 of Dijkstra's algorithm

Step 2: Visit the unvisited vertex with the shortest distance from the source so far, add it to the set of visited vertices, find the distances from the source through this vertex to each unvisited vertex, and replace any distances that are shortened (it will be 1 plus the new edge weight in this case):

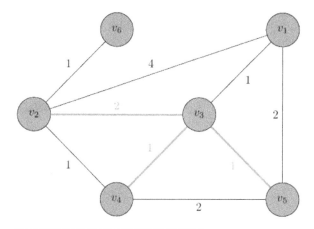

Vertex	Shortest Distance	Previous Vertex
v_1	0	
v_2	1 + 2 = 3	v_3
v_3	1	v_1
v_4	1 + 1 = 2	v_3
v_5	2	v_1
v_6	∞	

Visited Vertices	Unvisited Vertices
v_1, v_3	v_2, v_4, v_5, v_6

Figure 9.11 – Step 2 of Dijkstra's algorithm

Step 3: Visit the unvisited vertex with the shortest distance from the source so far (v_4), add it to the set of visited vertices, find the distances from the source through this vertex to each unvisited vertex, and replace any distances that are shortened.

This time, both v_4 and v_5 have distance 2, so we arbitrarily choose v_4:

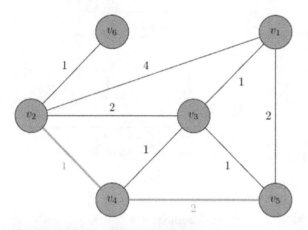

Vertex	Shortest Distance	Previous Vertex
v_1	0	
v_2	3	v_3
v_3	1	v_1
v_4	2	v_3
v_5	2	v_1
v_6	∞	

Visited Vertices	Unvisited Vertices
v_1, v_3, v_4	v_2, v_5, v_6

Figure 9.12 – Step 3 of Dijkstra's algorithm

Here, the distance to v_2 would be 2 + 1 = 3, which is not an improvement. The distance to v_5 would be 2 + 2 = 4, which is not an improvement. Therefore, this step makes no updates, and we will simply move on to the next smallest distance on the list.

Step 4: Visit the unvisited vertex with the shortest distance from the source so far (v_5), add it to the set of visited vertices, find the distances from the source through this vertex to each unvisited vertex, and replace any distances that are shortened.

There are no unvisited vertices adjacent to v_5, so we move on to the next step:

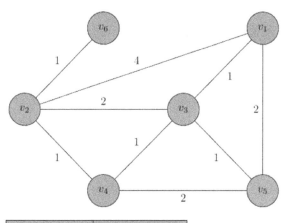

Figure 9.13 – Step 4 of Dijkstra's algorithm

Step 5: Visit the unvisited vertex with the shortest distance from the source so far (v_2), add it to the set of visited vertices, find the distances from the source through this vertex to each unvisited vertex, and replace any distances that are shortened:

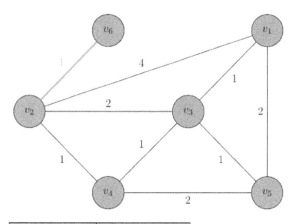

Figure 9.14 – Step 5 of Dijkstra's algorithm

Step 6: Visit the unvisited vertex with the shortest distance from the source so far (v_6), add it to the set of visited vertices, find the distances from the source through this vertex to each unvisited vertex, and replace any distances that are shortened:

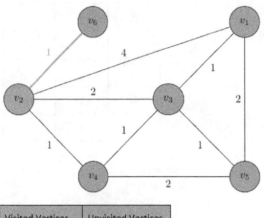

Vertex	Shortest Distance	Previous Vertex
v_1	0	
v_2	3	v_3
v_3	1	v_1
v_4	2	v_3
v_5	2	v_1
v_6	4	v_2

Visited Vertices	Unvisited Vertices
$v_1, v_2, v_3, v_4, v_5, v_6$	

Figure 9.15 – Step 6 of Dijkstra's algorithm

The unvisited set of vertices is now empty, and the shortest distance to v_2 is 3. The last edge of the shortest path is the edge from v_3 to v_2 as per the table. The previous vertex from v_3 of its shortest path is $v1$, so the shortest path the algorithm found from $v1$ to $v2$ is as follows:

$$v_1 - v_3 - v_2$$

This path and distance match what we found by brute force previously, but we followed a systematic algorithm. Let's also write down the extra, bonus results Dijkstra's algorithm gives us: the shortest paths from v_1 to every other node. We summarize the findings in the following figure:

Destination	Path	Distance
v_2	$v_2 \leftarrow v_3 \leftarrow v_1$	3
v_3	$v_3 \leftarrow v_1$	1
v_4	$v_4 \leftarrow v_3 \leftarrow v_1$	2
v_5	$v_5 \leftarrow v_1$	2
v_6	$v_6 \leftarrow v_2 \leftarrow v_3 \leftarrow v_1$	4

Figure 9.16 – The shortest paths from v_1 to every other vertex found with Dijkstra's algorithm

In this section, we have learned about Dijkstra's algorithm for finding the shortest paths between vertices in a network and worked through a small example by hand. Given the new understanding this example has given us, we will learn how to implement Dijkstra's algorithm in Python so that we can solve larger problems!

Python Implementation of Dijkstra's Algorithm

We have now learned how Dijkstra's algorithm works, but we will now implement it in Python.

The input to the algorithm will be a network and a source vertex. The simplest way we can represent a network is with a weight matrix like we introduced in *Chapter 8*, *Storage and Feature Extraction of Graphs, Trees, and Networks*. For the graph in *Figure 9.7*, we have the following weight matrix:

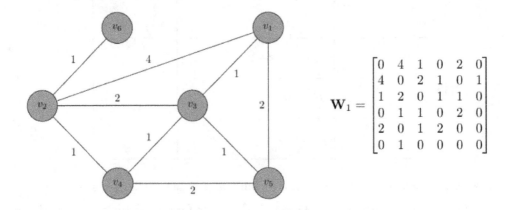

Figure 9.17 – A small network and its weight matrix

In the context of a shortest-distance problem, this weight matrix may be called a distance matrix, but we will refrain from using this terminology because, as we have seen in previous sections, these shortest path problems may or may not actually refer to distances.

The output from the algorithm will be a table like the one at the upper right of *Figure 9.15*, giving the shortest distance from the source vertex to each of the other vertices.

The table in *Figure 9.16* could be generated directly as well, but we will save this to be done outside the main function for Dijkstra's algorithm.

Let's write a function that takes the weight matrix and source vertex as an input, performs Dijkstra's algorithm, and returns the table. Since this code is a little long, we will display it in small parts and explain each step.

First, we will import NumPy and write some quick documentation. This just summarizes what the following code will do. This is the best practice when you write a new function or a significant batch of code:

```
import numpy

# Dijkstra's algorithm for finding shortest paths from the
    # source vertex to all other vertices in the graph
#
```

```
# INPUTS
# W - a weight matrix. It should be a square matrix
# i - the number of the source node
#
# OUTPUTS
# shortestDistances - the shortest distances from the source to
  # each vertex
# previousVertices - the previous vertex to the destination in
  # shortest path from the source to a destination
```

Second, we will define the function called Dijkstra, which will take a weight matrix, W, and a vertex, v_i, as the source. The first task we will do is find the number of vertices, initialize several NumPy arrays to store the data we will output for the table, which is the status of each vertex as unvisited or not.

We will also set the initial distances to the destinations as ∞, set the distance to the source vertex to 0, and mark the source vertex as a visited vertex:

```
def Dijkstra(W, i):
    # find the number of vertices
    n = W.shape[0]

    # initialize the shortest distances to infinity
    shortestDistances = numpy.array([numpy.inf] * n)

    # initialize the previous vertices
    previousVertices = numpy.array([numpy.inf] * n)

    # initialize the unvisited vertex set to be full
    unvisited = numpy.array([1] * n)

    # mark the source as visited
    unvisited[i - 1] = 0

    # initialize distance from the source to the source as 0
    shortestDistances[i - 1] = 0
```

Third, we will create a loop that will iterate once for each vertex. Within the loop, we find the nearest unvisited vertex, x, and visit it:

```python
# loop for iteration per vertex until the unvisited set is
#   empty
for _ in range(n):
    # find the distances to all unvisited adjacent vertices
    #   and set others to 0
    distances = shortestDistances * unvisited

    # find the index of the nearest unvisited vertex (where
    # distances > 0)
    x = numpy.argmin(numpy.ma.masked_where(
        distances == 0, distances))

    # mark vertex x as visited
    unvisited[x] = 0
```

Fourth, we will iterate over each vertex, and if any adjacent, unvisited vertices have their shortest distance from the source reduced by passing through the current vertex, we save this new shortest distance and save the current vertex as the vertex to visit prior to this destination in the shortest path located so far in the algorithm:

```python
# iterate through the vertices
for v in range(n):

    oldDistance = shortestDistances[v]
    newDistance = shortestDistances[x] + W[v,x]
    adjacent = W[v,x] > 0
    unvis = unvisited[v]

    # if v and x are connected, v has not been visited,
    #   and we find a shorter distance to node v...
    if adjacent and unvis and oldDistance > newDistance:
        # save the shortest distance found so far
        shortestDistances[v] = shortestDistances[x] + W[v,x]
```

```
        # save the previous vertex
        previousVertices[v] = x
```

Lastly, we will print a table just like we have at the upper right of *Figure 9.15*. Note that we add 1 to deal with the fact that Python numbers the vertices from 0 while we number them from 1. We also return the same information in the form that Python stores it by default in case we want to chain the algorithm to some more work:

```
    # print the table similar to the book
    print(numpy.array([numpy.arange(n) + 1, shortestDistances,
                       previousVertices + 1]).T)
    # return the outputs
    return shortestDistances, previousVertices
```

Now that we have written this implementation of Dijkstra's algorithm, we should try it out. Now, of course, it should work on large problems, but we recommend you always test out new code, especially long ones, on a problem with a known solution just to verify that it is working well.

Example – shortest paths

So, let's use the small network and weight matrix from *Figure 9.17* and see whether it will create the outputs we know are correct as shown in *Figure 9.15*.

First, we save the weight matrix as a NumPy array:

```
# Create a weight matrix for the network in Figure 9.15
W1 = numpy.array([[0, 4, 1, 0, 2, 0],
                  [4, 0, 2, 1, 0, 1],
                  [1, 2, 0, 1, 1, 0],
                  [0, 1, 1, 0, 2, 0],
                  [2, 0, 1, 2, 0, 0],
                  [0, 1, 0, 0, 0, 0]])
```

Then, we call Dijkstra's algorithm on the matrix, W_1, and source vertex, v_1:

```
# Run Dijkstra's algorithm with a source at vertex v1
Dijkstra(W1, 1)
```

The output is as follows:

```
[[ 1.  0. inf]
 [ 2.  3.  3.]
 [ 3.  1.  1.]
 [ 4.  2.  3.]
 [ 5.  2.  1.]
 [ 6.  4.  2.]]

(array([0., 3., 1., 2., 2., 4.]), array([inf, 2.,  0.,  2.,
  0.,  1.]))
```

The array that is outputted first is exactly the same as the table we found in *Figure 9.15* by applying Dijkstra's algorithm by hand for this problem.

The second part, which is what was actually returned by the function, is the same as the right two columns, just with the numbers less by 1 due to Python's preference to start counting from 0.

This chart is nice, but what about actual paths? It would be convenient if we could generate the paths themselves as we did by hand in *Figure 9.16*. This would be tedious with a large path, so let's write a short function to do that for us!

First, we define a new function and initialize some variables and lists:

```
# Use the previousVertices chart to construct the shortest path
  # from input source to input destination and print a
    # string showing the path

def printShortestPath(shortestDistances, previousVertices,
   source, destination):
    # avoid off-by-one error
    source -= 1
    destination -= 1

    # convert previousVertices to integers
```

```python
previousVertices = previousVertices.astype(int)

# initialize the path with the destination
path = [destination]
```

Next, we add the previous vertex from the table over and over until we reach the source:

```python
# add the previous vertex from previousVertices until we
  # reach the source
# the source
for _ in range(previousVertices.shape[0] - 1):
    # if the source is in the path, stop
    if path[-1] == source:
        break
    # if the source is not in the path, add the previous
      # vertex
    else:
        path.append(previousVertices[path[-1]])
```

Lastly, we create and print a string similar to the second column of the table in *Figure 9.16*:

```python
# initialize an output string
output = []

# iterate through the path backwards (source to
  # destination)
for i in numpy.flip(path):
    # construct a list of strings to output
    if i > 0:
        output.append('->')

    output.append('v' + str(i + 1))

# print the strings with no spaces
print('Path =', *output, '\t\t Distance =',
      shortestDistances[destination])
```

With this code written, let's run it to find short paths from v_1 to each other vertex:

```
for i in range(2,7):
    printShortestPath(shortestDistances, previousVertices, 1,
        i)
```

The output is as follows:

```
Path = v1 -> v3 -> v2              Distance = 3.0
Path = v1 -> v3                    Distance = 1.0
Path = v1 -> v3 -> v4              Distance = 2.0
Path = v1 -> v5                    Distance = 2.0
Path = v1 -> v3 -> v2 -> v6        Distance = 4.0
```

As you can see, it totally matches the table from *Figure 9.16*.

All looks great for this example, but let's look at an example with an extra difficulty: a network where some pairs of vertices have no path between them.

Example – A network that is not connected

Consider the following network and weight matrix:

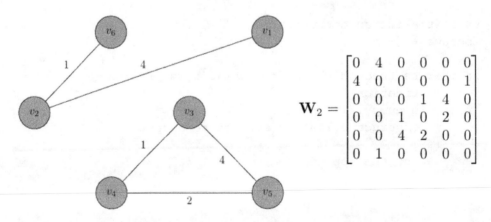

Figure 9.18 – A network that is not connected

This graph is broken down into two connected components that are not connected to one another by any edges, so there will be no shortest path between vertices in opposite components. As such, feeding this into Dijkstra's algorithm as we have written it cannot work in the same way. We will need to adapt our methods to find the shortest paths.

First, let's save the weight matrix as a NumPy array:

```
# Create a weight matrix for the network in Figure 9.16
W2 = numpy.array([[0, 4, 0, 0, 0, 0],
                  [4, 0, 0, 0, 0, 1],
                  [0, 0, 0, 1, 4, 0],
                  [0, 0, 1, 0, 2, 0],
                  [0, 0, 4, 2, 0, 0],
                  [0, 1, 0, 0, 0, 0]])
```

Next, let's write a small function to do a few things: (1) find all vertices connected to the source node using the `isConnected` function and (2) run Dijkstra's algorithm to find the shortest paths:

```
# find the shortest paths to connected vertices
def distancesWithinComponent(source):
    # initialize the connected component
    component = [source]

    # construct the connected component
    for i in range(1, W2.shape[0] + 1):
        if i != source and isConnected(W2, source, i):
            component.append(i)

    # find the weight matrix correponding to the connected
        # component
    subnetwork = W2[numpy.array(component) - 1,:][:,numpy.
     array(component) - 1]

    # run Dijkstra's algorithm
    return Dijkstra(subnetwork, 1)
```

Let's run it from vertex v_1:

```
distancesWithinComponent(1)
```

The output is as follows:

```
Vertex 1 and vertex 2 are adjacent
There is a path with 2 edges from vertex 1 to vertex 6
[[ 1.   0. inf]
 [ 2.   4.   1.]
 [ 3.   5.   2.]]
(array([0., 4., 5.]), array([inf,  0.,  1.]))
```

Note that the table is slightly off—the first column should be 1, 2, and 6. However, we have constructed a subnetwork and renumbered the vertices to 1, 2, and 3. But, clearly, we see the shortest path from v_1 to v_2 simply follows the edge connecting them for a length of 4 and the shortest path from v_1 to v_6 passes through v_2 with a length of 5.

Next, let's run it with a source in the other component, v_3:

```
distancesWithinComponent(3)
```

The output is as follows:

```
Vertex 3 and vertex 4 are adjacent
Vertex 3 and vertex 5 are adjacent
[[ 1.   0. inf]
 [ 2.   1.   1.]
 [ 3.   3.   2.]]
(array([0., 1., 3.]), array([inf,  0.,  1.]))
```

Here, the vertices are v_3, v_4, and v_5. The shortest path from v_3 to v_4 simply traverses the edge connecting them of length 1, while the shortest path from v_3 to v_4 passes through v_5 and has a length of 3. These results are fairly obvious for the small graph we used, but it is good to see that we can use the code we have written to work with disconnected networks.

Summary

In this chapter, we used our understanding of graph structures, including trees and networks, from *Chapter 8, Storage and Feature Extraction of Graphs, Trees, and Networks*, and learned about some practical graph-oriented problems and popular algorithms for solving them.

We began by learning about graph searches where we traverse a graph to discover its structure and perhaps do some calculations at each vertex. Then, we moved on to perhaps the most common graph search algorithm, DFS. We did an example on a small graph by hand before writing a Python implementation of the algorithm, which we confirmed led to the same results as the example we did by hand.

Then, we moved on to a very practical problem: finding the shortest paths between vertices in networks. This problem has applications in finding optimal travel routes, sending messages over a computer network through good paths, efficiently delivering electricity over electrical grids, and many other areas. With some networks, there are no paths between certain vertices, so we wrote a quick procedure to verify vertices are connected to each other by a path. Then, we used some counting methods we learned in *Chapter 4, Combinatorics Using SciPy*, to show that brute-force methods to finding shortest paths are infeasible.

In the next section, we introduced Dijkstra's practical algorithm for finding the shortest path from a source vertex to each other vertex in the network since brute-force methods were not effective. It is a greedy algorithm that takes the step that seems most advantageous at each iteration. We first carried out the algorithm step by step by hand on a small problem to build some understanding of how it works.

In the last section, we wrote a Python implementation of Dijkstra's algorithm from scratch that works just like the example we did by hand. It generated precisely the same optimal path for that example, but we also showed how it can immediately be applied to other problems by simply inputting the weight matrix and the source node.

Next, we will move on to *Part III* of the book, which focuses on real-world applications of the mathematics we have learned, including linear regression for machine learning, web searches with Google's PageRank algorithm, and principal components analysis, a method for dimensionality reduction that allows us to store large datasets with fewer variables.

Part III – Real-World Applications of Discrete Mathematics

Here you will learn how to apply discrete math to real-world, large-scale problems, including machine learning—in the shape of regression analysis for building predictive models and principal component analysis for dimensionality reduction—and modern web searches.

This part comprises the following chapters:

- *Chapter 10, Regression Analysis with NumPy*
- *Chapter 11, Web Searches with PageRank*
- *Chapter 12, Principal Component Analysis with Scikit-Learn*

10
Regression Analysis with NumPy and Scikit-Learn

The objective of this chapter is to predict an unknown variable based on samples of one or more other variables. In the simplest case, we have a sample of paired data $(x_1, y_1), ..., (x_n, y_n)$ and need to find a line that best fits the data (that is, a line that passes through or is close to most of the data points) with SciPy implementations of the least-squares regression model. We will then extend the method to fit nonlinear curves and to take whole databases $(x_{11}, x_{12}, ..., x_{1k}, y_1), ...,(x_{n1}, x_{n2}, ..., x_{nk}, y_n)$ and try to predict y based on k input variables.

We will also be using some Python libraries, such as SciPy, NumPy, and scikit-learn. SciPy is an open source Python library for scientific computing, and NumPy will help us to work with multidimensional arrays and matrices and apply high-level mathematical functions to these arrays. Scikit-learn is a machine learning library, and we will be using the regression classes that come with it.

By the end of this chapter, you should have learned about the theory behind regression (such as the line of best fit, the least-squares method, and more) and how to implement this theory for real-world datasets to make predictions.

In this chapter, we will be covering the following topics:

- Best-fit lines and the least-squares method
- Least-squares lines with NumPy
- Least-squares curves with SciPy and NumPy
- Least-squares surface with SciPy and NumPy

> **Important Note**
> Please navigate to the graphic bundle link to refer to the color images for this chapter.

Dataset

For this chapter, we will be using a dataset that contains technical specifications for different cars. This dataset is a modified version of the `MPG_dataset.csv` available here: `https://www.kaggle.com/uciml/autompg-dataset`. Some of the columns of the original dataset were removed since they are not relevant to this chapter.

The **columns** of the dataset are as follows:

- **mpg**: Miles per gallon (continuous variable)
- **cylinders**: Number of cylinders in the car (multi-valued discrete variable)
- **displacement**: Combined volume of all the cylinders (continuous variable)
- **horsepower**: Unit of power (continuous variable) – *target/dependent variable*
- **weight**: Weight of the car (continuous variable)
- **acceleration**: Acceleration of the car (continuous variable)

Let's say that we are trying to buy a car and our deciding factor is horsepower. However, we have access to values for all other variables (mpg, displacement, weight, and acceleration) except for horsepower. Here are some questions we will try to answer in this chapter:

- Is there any relationship between the horsepower and weight of a car?
- If there is a relationship, how strong a relationship is it? Is it a linear relationship?
- Is there any way for us to predict what will the horsepower value be given any one or more of the other variables?

Linear regression can answer the preceding questions. We will learn some general concepts about linear regression and then use this dataset to answer the questions just posed.

Here we can see some pair plot code showing the relationship between the different variables present in our dataset. The plot gives a general idea about how our variables are related to each other:

```
import numpy as np
import pandas as pd
import seaborn as sns
import matplotlib.pyplot as plt

#Importing the csv file
df = pd.read_csv("auto_dataset.csv",index_col=0)

#Plotting the pairplot
sns.pairplot(df, diag_kind="kde")
plt.show()
```

The output of the code is shown here:

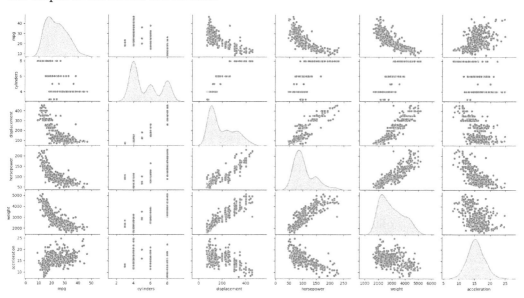

Figure 10.1 – Pair plot showing the relationship between different variables in the dataset

You can see that some of the plots have a straight-line relationship (linear relationship) while others do not. There can be different kinds of relationships between variables, such as logarithmic, exponential, and polynomial. The plots along the diagonal of the figure show the distribution of the variables.

Next, we will discuss best-fit lines and the least-squares method, which will aid our understanding of regression analysis.

Best-fit lines and the least-squares method

In this section, we will learn about best-fit lines and the least-squares method and see some very simple plots to enforce these concepts. Best-fit lines will be used to fit a dataset, and through doing this, it is made sure that the least-squares error is minimized.

Variable

A variable is a feature that does not have a fixed value. For example, let's say you want to buy a car and the most important feature for you is horsepower. You then go ahead and look at different car models to compare the horsepower values. Here the horsepower is a variable, since it takes different values based on the car model.

Linear relationship

Linear relationship is a term used to describe a straight-line relationship. This can be expressed mathematically as the equation of a line.

For example, if you are interested in finding the relationship between two variables, such as *horsepower* and *weight*, for different cars, and if the *horsepower* increases or decreases linearly (straight-line relationship) with *weight*, then we can say that these variables have a linear relationship.

Regression

In a typical regression problem, we try to predict the value of an output variable (dependent variable) given some input variable (independent variable), based on some examples of input data points that we have outputs for. A simple **linear** regression problem can be represented mathematically as shown here (this is the equation for a line):

$$Y \approx \beta_o X_o + \beta_1 X_1 = Y$$

Here, we are trying to find the value of Y based on our knowledge of independent/predictor variable X_1, also called features. β_0 and β_1 are unknown constants (called model parameters) that we will have to figure out based on the example data points, which will be represented in the form of ordered pairs (X_i, Y_i), where $i = 0,1,2,....,n$. X_0 is always equal to 1. An independent variable is something whose value we can change as we want and see the changes in the dependent variable (here, Y).

It is important to keep in mind that not all X_i can be mapped to all Y_i perfectly or generalize to new outputs, but we try to get as close to the ideal match (perfect match) as possible, hence the symbol $\approx$ to convey the fact that it is an approximate model. The approximate value of the prediction is represented by $\hat{Y}$.

An appropriate question to ask here would be, why are we trying to predict a value of Y? Often, in real-life scenarios, we would have the values for X (the independent variable) and Y (the dependent variable). However, if we want to predict the value of Y for a certain X that is of interest to us, we will want to have access to an equation like the previous one.

For example, say we want to buy a car and we know that there is a linear relationship between weight and horsepower. We have some historical data that has weights of different cars as well as their corresponding horsepower. Let's say that we come across the weight of a car but see no mention of the horsepower. Hence, we can use our data to predict what the horsepower might be for this car for which we only know the weight.

The equation shown previously requires two variables for us to fit a linear regression model. However, we can do the same for more variables/for higher dimensions, the equation for which is shown here (equation for a plane/surface):

$$Y \approx \beta_0 + \beta_1 X_1 + \beta_2 X_2 + \beta_3 X_3 + ... + \beta_n X_n = \hat{Y}$$

We can write the preceding equation in a vectorized form as shown here:

$$\hat{Y} = \beta . X$$
$$\hat{Y} = \beta^T X$$

Here, the following applies:

- β is the model parameter vector containing $\beta_o, \ldots, \beta_n$ – these are the parameters we will change in order to get Y and $\hat{Y}$ as close to each other as possible.
- X is the model feature vector containing $X_o, X_1, \ldots, X_n$.
- We are doing a dot product between β and X, which will result in $\beta_0 + \beta_1 X_1 + \beta_2 X_2 + \beta_3 X_3 + \ldots + \beta_n X_n$.

Both β and X are column vectors. With our knowledge of the vectorized form of an equation, we will now move on to learn about the line of best fit.

The line of best fit

The line of best fit, also called a trendline, is an educated guess regarding what the linear relationship between the independent and dependent variables should be; it is the equation of a line that best fits the given data. Let's try to understand this with the help of an example. We will plot some X and Y values and see how the line of best fit varies depending on the data points. Here, the R^2 values give us a measure of the strength of the linear relationship between the variables. An R^2 value of 1 suggests that the variables are indeed linearly related. The lower the R^2 value, the more likely it is that the variables might not be linearly related:

X	Y
1	1
2	2
3	3
4	4
5	5
6	6

Figure 10.2 – Data points used for a line of best fit that passes through all points

We will now plot the preceding data points and draw a trendline through the points:

Best-fit lines and the least-squares method 241

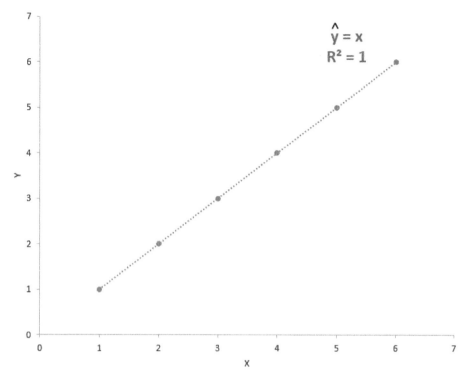

Figure 10.3 – A line of best fit that passes through all data points

In the preceding plot, we see that the line of best fit passes through all the data points, and hence we obtain the equation $\hat{Y} = X$. However, this does not always have to be the case, since the X (predictor) and Y (dependent) values are usually not the same in real-world examples.

Next, we will try to plot a similar graph but with mismatched X and Y to see how the best-fit line as well as the equation for the line changes:

X	Y
1	3
2	2
3	4
4	6
5	8
6	7

Figure 10.4 – Data points used for a line of best fit that does not pass through all points

We will now plot these data points and draw a trendline through the points and compare the difference between the previous plot and this one. The following graph shows the plotting of the points from the preceding table:

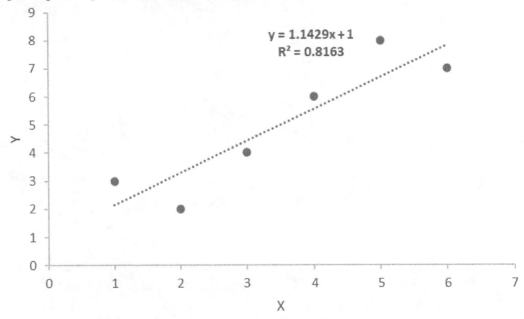

Figure 10.5 – A line of best fit that does not pass through all the data points

We can see from *Figure 10.5* that the best-fit line does not pass through all the points and hence the equation of the best-fit line has an intercept ($\beta_0 = 1$) and a slope ($\beta_1 = 1.1429$).

It is important to keep in mind that even if you can draw a best-fit line through some data points, it does not mean the variables (X and Y) have a linear relationship. In such cases, it is always a good idea to make a scatterplot for the data points and look at the plot to see whether a linear relationship makes sense for that dataset. One such example is shown here:

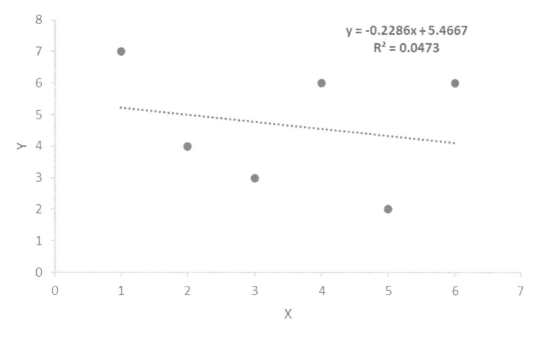

Figure 10.6 – A line of best fit for variables that are not linearly related to each other

The preceding figure shows an example where the line of best fit for a linear relationship is not a good idea. Hence, non-linear lines of best fit such as polynomial functions could be a better idea.

Now that we have seen what a line of best fit is, the next step is to figure out how these lines are constructed; this will be covered in the next section.

The least-squares method and the sum of squared errors

The sum of squared differences between the value of actual Y and the predicted $\hat{Y}$ is called the **sum of squared errors** (**SSE**). If the data points (actual and predicted) are identical, then the SSE is 0. This is also a measure of the variance: the greater the variance, the greater the SSE, and vice versa. In an ideal case, we would want the SSE to be small, and the best-fit line helps us to achieve this goal. The SSE can be mathematically represented as follows:

$SSE = e_1^2 + e_2^2 + \ldots + e_n^2$

Here $e_1, e_2, ..., e_n$ are the differences between the $\hat{Y}$ (predicted) and Y (actual), also called **residuals**. The following figure shows how the predicted Y values differ from actual Y values, hence giving rise to residuals:

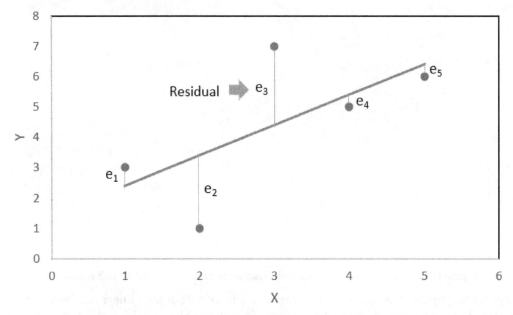

Figure 10.7 – Residuals

The preceding figure shows the residuals, in other words, the distance of the actual Y from the line of best fit (or predicted $\hat{Y}$). The least-squares approach tries to minimize the SSE by choosing the values of β parameters. To find the value of β that minimizes the SSE, there is a closed-form equation that can be used to get the result directly. This equation is called the normal equation and is stated here:

$$\hat{\beta} = (X^T X)^{-1} X^T Y$$

Here, the following applies:

- $\hat{\beta}$ is the value of the parameters that minimize the SSE.
- Y is the vector of target values ranging from $Y_1, ..., Y_n$.

The normal equation must compute the inverse of $X^T X$, which is an $(n+1) \times (n+1)$ matrix (since we have n feature variables and $X_0 = 1$). The computational complexity of such an inversion is of the order $O(n^3)$, depending on the implementation. Hence, if we have double the number of features, then the computational time gets multiplied by $2^3 = 8$ times.

Another thing to keep in mind is that X^TX might not be invertible in all cases.

Now that we have an idea about line of best fit, the least-squares error, and their mathematical formulations, we will now learn how to apply these to examples using Python.

Least-squares lines with NumPy

In this section, we will learn how to fit a line to a dataset by using the normal equation as well as by using Python libraries. We will also find the parameter values (β) and use these values to predict the Y values for some X value of our choice.

The relationship between the variables (horsepower and weight) can be represented by the following mathematical formulation:

$$Y \approx \beta_o + \beta_1 X$$

Our goal is to find the values for β_o and β_1. Here, horsepower is the dependent variable (Y) and weight is the independent variable (X).

Before beginning the coding part, make sure that the Python file that you are editing and `auto_dataset.csv` are in the same folder. If not, make sure to include the path for the `.csv` file location in the Python file so that it can be read and used for computations. Also, the packages used in the coding exercises (`numpy`, `pandas`, `seaborn`, `matplotlib.pyplot`, and `sklearn`) should be installed to avoid error messages. These packages can be installed by typing `pip install numpy` (or whatever package you want to install) in the terminal.

We will begin by importing all the required packages. This can be done using the following block of code:

```
Import numpy as np
import pandas as pd
import seaborn as sns
import matplotlib.pyplot as plt
```

Next, we will read the CSV file and import the data to the Python workspace and check the shape of the data frame:

```
df = pd.read_csv("auto_dataset.csv")
df.shape
```

The output is this:

```
(392, 7)
```

Next, we will use the normal equation to find the parameter values, which will then be used for prediction purposes. Here, weight is chosen as the X value and horsepower is the Y value:

```
X = df["weight"]
Y = df["horsepower"]

X_b = np.c_[np.ones((392,1)),X] #here we are adding X_o = 1 to
  all the feature values
 beta_values = np.linalg.inv(X_b.T.dot(X_b)).dot(X_b.T).dot(Y)
 print(beta_values)
```

The output is this:

```
array([-12.1834847 ,    0.03917702])
```

We found the value of β_o is -12.1834847 and β_1 is 0.03917702.

The equation for the best-fit line is *horsepower = -12.183 + 0.0392 * weight*.

Let's try to predict the value of horsepower for a car given that we know what its weight is. We will try to predict the horsepower values for the cars for which we already know these values (weight of car = 2500 and 2045) and hence compare the actual and predicted values. We will use equation 4 and the β values obtained in the previous step to find the predicted values for horsepower:

```
X_new = np.array([[2500],[2045]])
X_new_b = np.c_[np.ones((2,1)),X_new]
y_predict = X_new_b.dot(beta_values)
print(f"Weight of car = 2500; predicted horsepower is
   {y_predict[0]:.3}; actual horsepower is 88")
print(f"Weight of car = 2045; predicted horsepower is
   {y_predict[1]:.3}; actual horsepower is 68")
```

The output is this:

```
Weight of car = 2500; predicted horsepower is 85.8; actual
   horsepower is 88
Weight of car = 2045; predicted horsepower is 67.9; actual
   horsepower is 68
```

Hence, we were able to predict the horsepower of the cars for which we knew the weights. We can see that the predicted and actual values are close enough but not the same. This happens because this is an approximation since the best-fit line does not pass through all the data points and we minimized the SSE.

Now that we have seen how we can use the normal equation to find β values, we will plot all the data points and the trendline. We will use the parameter values obtained previously to construct the equation for a line that will help us predict the `horsepower` values for cars for which we know the weights:

```
X_plot= np.array([[1500],[6000]])
X_plot_b = np.c_[np.ones((2,1)),X_plot]
Y_plot = X_plot_b.dot(beta_values)

Equationline = "Y ={:.3f}+{:.3f}X".format(beta_values[0], beta_values[1])
plt.plot(X_plot, Y_plot, "r-", label = Equationline)
sns.scatterplot(X,Y, label = "Training Data")
plt.legend()
plt.show()
```

The output of the code is shown here:

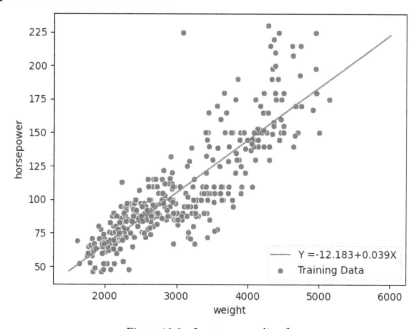

Figure 10.8 – Least-squares line fit

From the preceding plot, we see that vehicles with higher weights tend to have higher horsepower and vice versa. The general trend of the plot is that as the weight of the vehicle increases, so does the horsepower.

Let's repeat the preceding linear regression using scikit-learn. More information about this package and its APIs can be found here: `https://scikit-learn.org/stable/user_guide.html#user-guide`. Information regarding ordinary least squares regression can be found here: `https://scikit-learn.org/stable/modules/generated/sklearn.linear_model.LinearRegression.html#sklearn.linear_model.LinearRegression`.

We will obtain the same β values but with just a few lines of code using scikit-learn:

```
import pandas as pd
import numpy as np
from sklearn.linear_model import LinearRegression
reg = LinearRegression()
df = pd.read_csv("auto_dataset.csv")
X = df["weight"]
Y = df["horsepower"]
X = X.values.reshape(-1,1)
Y = Y.values.reshape(-1,1)

reg.fit(X, Y)
print("The value obtained for beta_o is: ", reg.intercept_)
print("The value obtained for beta_1 is: ",reg.coef_)
```

```
The value obtained for beta_o is:    [-12.1834847]
The value obtained for beta_1 is:    [[0.03917702]]
```

We can also use scikit-learn for prediction, and this can be done in just one line of code, as follows. We will use the same weights of 2500 and 2045 as in the previous example and hence obtain the same predicted `horsepower` values:

```
X_new = np.array([[2500],[2045]])
print(reg.predict(X_new))
```

```
[[85.75906307]
 [67.93351937]]
```

In this section, we learned about how to obtain the best-fit line for a dataset using Python and some of its packages, such as NumPy and scikit-learn. We also learned about how to predict a certain value *Y* given the predictor variable *X*.

In the next section, we will learn about fitting least-squares curves to a dataset. This applies to cases where the *X* and *Y* variables do not have a linear relationship.

Least-squares curves with NumPy and SciPy

We will now learn how to fit curves to a dataset. For this section, we will investigate the relationship between `horsepower` and `mpg` for a vehicle. From *Figure 10.1*, we know that the relationship between these two variables is not linear; hence, we will use power 2 of our feature variable *X* as an input to the model. This is called polynomial regression. Here, we are using a linear model to fit a non-linear dataset.

Here's how we will import the required Python packages and select the *X* and *Y* of interest from the pandas data frame, `df`:

```
import numpy as np
import pandas as pd
import seaborn as sns
import matplotlib.pyplot as plt
from sklearn.preprocessing import PolynomialFeatures
from sklearn.linear_model import LinearRegression

#Importing the dataset as a pandas dataframe
df = pd.read_csv("auto_dataset.csv")

#Selecting the variables of interest
X = df["horsepower"]
y = df["mpg"]

#Converting the series to a column matrix
X_new = X.values.reshape(-1,1)
y_new = y.values.reshape(-1,1)
```

We will be using scikit-learn's `PolynomialFeatures` class. For more information about this class, refer to this link: https://scikit-learn.org/stable/modules/generated/sklearn.preprocessing.PolynomialFeatures.html#sklearn.preprocessing.PolynomialFeatures. This will help us to transform our input data by adding a new feature to the dataset – the square of X (`horsepower`), which is a second-degree polynomial. We will use a polynomial of degree 2 of the form $Y = \beta_0 + \beta_1 X + \beta_2 X^2$:

```
poly_features = PolynomialFeatures(degree=2, include_
   bias=False)
X_poly = poly_features.fit_transform(X_new)
```

`X_poly` contains the original feature (X = *horsepower*) plus the squared value of the feature. We will now use the linear regression model as shown in the previous section, to carry forward our analysis:

```
reg = LinearRegression()
reg.fit(X_poly, y_new)
print("Y ={:.4f} X^2 {:.3f} X + {:.3f}".format(reg.coef_[0,1],
   reg.coef_[0,0], reg.intercept_[0]))
```

The output is as follows:

```
Y =0.0012 X^2 -0.466 X + 56.900
```

Hence, we have the equation for the best-fit curve, as shown in the preceding output.

Next, we will use our knowledge of this equation to plot the best-fit line and lay it on the top of a scatterplot of the actual data. We will vary the X values between the minimum and maximum `horsepower` values that are present in the dataset for plotting the best-fit curve and hence calculate the corresponding Y values using the equation obtained in the preceding step:

```
start = df["horsepower"].values.min()
stop = df["horsepower"].values.max()
X_plot = np.linspace(start, stop, 1000)
Y_plot = reg.coef_[0,1] * X_plot * X_plot + reg.coef_[0,0] *
X_plot + reg.intercept_[0]

Equationline = "Y ={:.4f} $X^2$ {:.3f} $X$ + {:.3f}".
   format(reg.coef_[0,1], reg.coef_[0,0], reg.intercept_[0])

sns.scatterplot(X,y, label = "Training Data")
plt.plot(X_plot, Y_plot, "r-", label = Equationline)
```

```
plt.legend()
plt.show()
```

The output of the code is shown here:

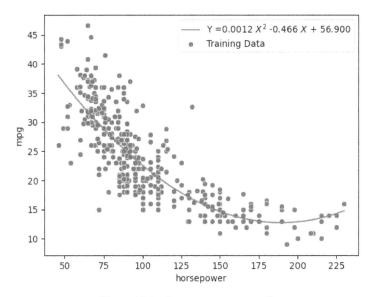

Figure 10.9 – Least-squares curve fit

The best-fit line for the dataset ($X = horsepower$; $Y = mpg$) is shown here. This line was overlayed on top of actual data points to show that the best-fit line is an approximation. When we try to predict the Y value for our choice of X, the algorithm will use the equation obtained to find us an approximate Y value. It is important to keep in mind that the predicted values are less reliable if we are trying to extrapolate outside the range of X values for the dataset.

In this section, we learned about how to fit least-squares curves to a non-linear dataset. To do so, we added a new feature variable equal to square of X. We then used the linear regression class provided by scikit-learn to find the β parameters. These parameters were then used to draw the best-fit curve. It is important to keep in mind that we might run into overfitting issues when fitting curves, which means that the prediction made using the equation we came up with might not be accurate.

Now that we know best-fit curves, the next practical step would be to learn about fitting least-squares surfaces.

Least-squares surfaces with NumPy and SciPy

An appropriate question to ask in this section would be to ask, "Why do we need to fit surfaces to a dataset?" It is important since 2D plots are not enough to show the relationship between the predictor variables (X_1, X_2, ..., X_n) and the predicted variable Y. In many real-life scenarios, Y is affected by more than one X variable, and hence to capture such a relationship, we would need a surface plot (3D), which can show the relationship between X_1, X_2, and Y. This relationship between the variables can be represented by the following mathematical formulation:

$$Y \approx \beta_o + \beta_1 X_1 + \beta_2 X_2$$

Our goal is to find the values for β_o, β_1, and β_2.

For this section, we will use the `horsepower` and `weight` values of a car as input for X_1 and X_2 respectively. The output variable will be displacement (Y). We can mathematically write this as follows:

$$Y \approx \beta_o + (\beta_1 * horsepower) + (\beta_2 * weight)$$

Here's how we will import the required Python packages and select the X and Y of interest from the pandas dataframe `df`:

```
from sklearn.linear_model import LinearRegression
import pandas as pd
import numpy as np
from mpl_toolkits import mplot3d
import matplotlib.pyplot as plt

#Importing the csv file and choosing the X and Y variables
df = pd.read_csv("auto_dataset.csv")
Y = df["displacement"]
X = df[["horsepower","weight"]]
```

Next, we will use the scikit-learn linear regression model to fit the X and Y values. We will print the values of the regression coefficients for our reference:

```
#Fitting the linear regression model
reg = LinearRegression()
reg.fit(X, Y)

# Printing the parameter values obtained after fitting the
    # model
print("The value obtained for beta_o is: ", reg.intercept_)
print("The value obtained for beta_1 and beta_2 are: ",reg.
    coef_[0] , "and", reg.coef_[1] )
```

```
The value obtained for beta_o is:   -135.95073526530456
The value obtained for beta_1 and beta_2 are:
  0.9757143655155813 and 0.07671670340152593
```

Hence, we can write the equation as follows:

$$Y \approx -135.951 + (0.976 * horsepower) + (0.0767 * weight)$$

Now that we have the model and the coefficients, the next step would be to plot the dataset as well as the model obtained (surface) to better understand the process. We will need to find the minimum and maximum values for horsepower and weight and then obtain 100 equally spaced values between the two values. Once we have these 100 equally spaced values, we can then use the preceding equation to obtain the corresponding Y values that will be used to make the surface plot. In addition, we will also plot the actual dataset that was used for obtaining the β values for comparison:

```
# Plotting the surface plot
X1_min = df["horsepower"].values.min()
X1_max = df["horsepower"].values.max()
X1_values = np.linspace(X1_min, X1_max, 100)

X2_min = df["weight"].values.min()
X2_max = df["weight"].values.max()
X2_values = np.linspace(X2_min, X2_max, 100)

Y_reg = reg.intercept_ + (reg.coef_[0] * X1_values) + (reg.
  coef_[1] * X2_values)
Y_plot = Y_reg.reshape(-1,1)

fig = plt.figure()
ax = fig.add_subplot(111, projection='3d')
ax.scatter(X.horsepower, X.weight, Y, color="red", s=1)
X1_plot, X2_plot = np.meshgrid(X1_values, X2_values)
surf = ax.plot_wireframe(X1_plot, X2_plot, Y_plot, rstride=10,
  cstride=10)
ax.view_init(50, 150)
ax.set_xlabel('Horsepower')
ax.set_ylabel('Weight')
ax.set_zlabel('Displacement')
plt.legend()
plt.show()
```

The output of the code is shown here:

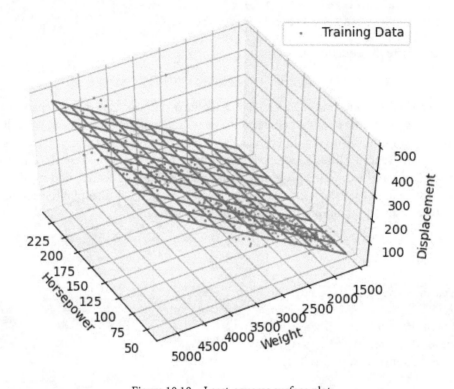

Figure 10.10 – Least-squares surface plot

The least-squares surface for the dataset (X = horsepower, weight; Y = displacement) is shown in the preceding figure. This surface was overlayed on top of actual data points to show that the least-squares surface is an approximation. When we try to predict the Y value for our choice of X values, the algorithm will use the equation obtained to find us an approximate Y value. It is important to keep in mind that the predicted values are less reliable if we are trying to extrapolate outside the range of X values for a dataset.

In this section, we learned about least-squares surfaces and how to implement them on a real-world dataset by using Python packages such as scikit-learn. Scikit-learn has a lot of other important classes that can be used for machine learning problems; it is always a good idea to go through the documentation at `https://scikit-learn.org/stable/modules/classes.html`.

Summary

In this chapter, we learned about regression, the least-squares method, and line, curve, and surface fitting. We also learned about how to apply these methods to a real-world dataset and how to predict the values for an output variable (Y) given access to some historical dataset that has both X and Y values. Caution should be taken if we are trying to extrapolate outside the range of X values for a dataset; the predicted values might not be reliable. You should now be able to apply these concepts to your own datasets and use Python libraries such as SciPy, NumPy, and scikit-learn to carry out regression analysis and prediction.

In the next chapter, we will learn about web searches from both mathematical and practical perspectives. We will also look at Google's PageRank algorithm and discuss the linear algebra involved.

11
Web Searches with PageRank

Searching the web is one of the first things we learn to do on the internet. The purpose, simply, is to find information of a topic of interest, but how does Google, or other search engines, take the words we search and effectively return what we want? This is the question we aim to answer in this chapter.

More specifically, this chapter discusses web searches from both a mathematical and practical perspective. We will first build the mathematical setting for common methods for web searches. We'll then look more deeply at Google's PageRank method and the linear algebra required. We'll then construct an implementation of PageRank that combines this linear algebra with the probabilistic aspects of PageRank we discussed in *Chapter 5, Elements of Discrete Probability*.

In this chapter, we will cover the following topics:

- The development of search engines over time
- How Google's PageRank algorithm works
- Implementing the PageRank algorithm in Python
- Applying the PageRank implementation to real data

By the end of this chapter, you will have learned how PageRank works, the linear algebra basis for it, why it is so effective, and how to implement the algorithm and apply it to real-world data.

> **Important Note**
> Please navigate to the graphic bundle link to refer to the color images for this chapter.

The Development of Search Engines over time

In this section, we will learn about the development of modern search engines on the internet. This will set the stage to learn about Google's PageRank algorithm. But, before we do that, let's briefly learn how older search engines worked and their shortcomings so that we can see why we need to tap into some deeper mathematics to solve the problem of ranking websites based on searches.

In the early 1990s, search engines were relatively simple. The search engine companies maintained databases of as many websites as they could. Users would search a word, say, `chicken`, and the search engines would search for websites using the word `chicken` and rank them based on how many times the word `chicken` appeared on the website. As you might suspect, this isn't necessarily the best approach.

There are several problems with these simple methods:

- Web pages where a certain search word occurs frequently are not necessarily what people are seeking when they do a web search. An FDA agricultural report might say `chicken` dozens of times, but not many users are likely to want agricultural reports.
- It automatically favored web pages with long passages of text, which were more likely to have more occurrences of `chicken`.
- There was little natural language processing, so a search of `chicken` might not return websites with `chickens` or `chicks`.
- Unpopular, little-used web pages with the word `chicken` were just as likely to show up in search results as popular websites that many users visit.
- It was easy to game the system: unscrupulous webmasters would add huge passages of transparent text full of commonly searched words such as `chicken` written hundreds of times or store such words in large passages of metadata just to drive traffic to their website.

Through the 90s, a bit more diversity in methods proliferated. One innovation was to allow searches with Boolean functions such as AND, OR, and NOT—so, you could search the following:

- `chicken AND sandwich` returns web pages with both words that hungry users may be seeking.
- `chicken OR rooster` returns web pages with either word that users interested in animals may be seeking.
- `chicken AND NOT egg` returns web pages with the word `chicken` but not the word `egg` to filter out web pages related to eggs from the web pages mentioning chickens.

In addition, some search engines introduced fuzzy logic, which could return web pages that are relevant to the search but not strictly satisfying the search. For example, a search of `chicken` may return web pages with the word `chicken` as well as web pages with the words `chicken`, `chick`, `chicks`, or even `nuggets`, `wings`, or `poultry`.

These innovations improved the quality of web searches, but they were still not nearly as effective as today's search engines, which seem to have a knack for returning the web pages you actually want.

This is not meant to be a comprehensive description of search engines in the mid-to-early 1990s, as there were some other algorithms used by the many search engines of the time—Yahoo, Lycos, Excite, and the like. But this should give you an understanding that the relatively simplistic search algorithms of the time period had many challenges.

To make things worse, the internet was growing exponentially, meaning web searches began returning hundreds, thousands, even millions of web pages. If search engines were returning so many web pages without using some more reasonable criteria for deciding how relevant or how important certain web pages were to move them to the top of the list when users searched, they were not very likely to return the right web pages without requiring users to sift through pages and pages of search results.

These failings prompted the need for a different kind of ranking method to move the "best" web pages to the top of the list when users search. With this in mind, we will continue to see how modern ranking algorithms, PageRank in particular, use linear algebra and probability to find the importance of web pages on the basis of which other websites link to them.

Google PageRank II

In this section, we will continue learning about Google's PageRank algorithm, which we started to look at in *Chapter 5, Elements in Discrete Probability*. As we discussed in that chapter, two students at Stanford University and later founders of Google, Larry Page and Sergey Brin, along with some researchers at Stanford, Rajeev Motwani and Terry Winograd, tapped into some existing academic literature on information retrieval in linked documents and merged several innovations to adapt the ideas for use in web searches.

The algorithm they developed, PageRank, was so effective that Google soon became totally dominant in the field of search engines in the late 1990s to early 2000s. This innovative PageRank algorithm still forms a part of Google's searching methods, although their methods have, of course, progressed significantly in the past 20 years by implementing information from user histories, user location, and the like in determining which websites are most likely relevant to users.

Without further ado, let's dive into learning just how PageRank works!

In a basic search engine using the PageRank algorithm, a user searches some terms and all the web pages with the terms, and perhaps web pages matching according to some fuzzy criteria, are returned. Then, the PageRanks of all the websites are found and sorted from highest to lowest. Finally, the results are displayed to the user in this descending order. Ideally, this means the most relevant websites will be shown to the user first.

To see how PageRank works under the hood, let's first quickly review what we learned about the PageRank algorithm in *Chapter 5, Elements of Discrete Probability*. Suppose we have an "internet" made up of a set of web pages. We will call the "internet" I and assume it has some finite number, N, of distinct web pages. In the real internet, this N numbers in the billions! We will refer to these web pages as follows:

$$I = \{W_1, W_2, \ldots, W_N\}$$

On I, we define two functions:

- Outgoing links, $C: I \to \{0, 1, 2, \ldots, N-1\}$, where $C(W_j)$ is the number of links leaving the jth web page, where self-links do not count and multiple links to the same web page count as a single link.
- PageRank, $PR: I \to [0,1]$, where we have $PR(W_j)$. It is calculated as follows:

$$PR(W_j) = \frac{1-d}{N} + d \sum_{W_i \in M(W_j)} \frac{PR(W_i)}{C(W_i)},$$

Here, $M(W_j)$ is the set of web pages linking to W_j. In other words, PageRank is $(1 - d)/N$ plus d times the sum of ratios of PageRank to outgoing links for each other web page linking to W_j. The constant $d \in (0,1)$ is called the damping factor. The authors set $d = 0.85$ in their original paper, although Google may have adjusted it since then. Regardless of the value of d, it can be shown that the function PR is a probability mass function, assigning probabilities to $W_1, W_2, ..., W_N$. Note that, by definition, the probabilities assigned by a probability mass function sum to 1.

> **Important Note**
>
> Note that there is some confusion in the literature about the first term of the PR calculation: sometimes the N is left out of the denominator. This does not have an important practical impact since we simply rank websites in descending order based on the outputs. However, the resulting PageRanks do not form a probability mass function without this N and so it is mathematically not quite so clean.

The idea behind PageRank is to take an imaginary "person" navigating this "internet" who will randomly click links and will eventually stop on a certain web page. The value d represents the probability that this person will click the next link at each step. The PageRank of a web page, $PR(W_i)$, represents the probability that this randomly clicking surfer will stop on web page W_i.

The PageRank algorithm initializes all the PageRanks to be equal, meaning they will initially be $1/N$ since they must add up to 1 in order for them to make up a probability distribution. Then, the PageRank algorithm will redo this calculation for each web page periodically to update the PageRank of each website based on changes in link structure and traffic patterns over time.

> **Important Note**
>
> In many implementations of the PageRank algorithm, it will actually compute the formulas over and over until all the PageRanks converge to a steady state where further calculations of the formula result in the same outputs, which tends to happen. When patterns in the links change, it is possible to carry out this procedure again to find the new PageRanks.

In *Chapter 5, Elements of Discrete Probability*, we created a small "internet" of just five web pages with a fixed linking structure. We have replicated the figure of this small "internet" represented as a directed graph in *Figure 11.1*, but with three changes:

- We have color-coded the vertices representing web pages and edges representing links between web pages.
- We have separated bidirectional arrows into separate arrows when two web pages each have links to the opposite web page.
- We added weights to the edges representing the amount of the PageRank of a source website, W_i, will be passed on to the web pages they link: that is, simply the following:

$$\frac{1}{C(W_i)}$$

This weight has a practical impact: if the web page simply links a huge number of other web pages, the full PageRank of the web page with the links will not have such a big impact on the rest of the "internet." This prevents a strategy for gaming the system: without this, web pages would try to have themselves added to giant web page indexes, which could increase their PageRank for a reason that probably has little to do with how relevant the web page is to the user's search.

This makes our diagram a directed *network* rather than a directed graph, as we see in the following figure:

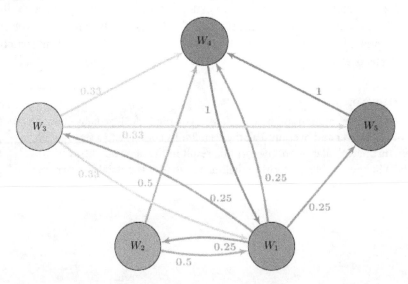

Figure 11.1 – A directed network representing an "internet" of five web pages, their links, and the proportion of PageRank carried over each link. All web pages, links, and weights are color-coded

In *Chapter 5, Elements of Discrete Probability*, we initialized the PageRanks to $1/5 = 0.2$ and used the linking structure to compute the PageRanks using the preceding functions for one iteration. The obtained PageRank are shown as follows:

$$PR(W_1) = 0.34$$

$$PR(W_2) = 0.07$$

$$PR(W_3) = 0.07$$

$$PR(W_4) = 0.38$$

$$PR(W_5) = 0.13$$

But we did this calculation by hand, which is not ideal, so we would like to replicate this calculation with matrix arithmetic.

As we learned in *Chapter 8, Storage and Feature Extraction of Graphs, Trees, and Networks*, the directed network of our small "internet" can be represented as a matrix:

$$\mathbf{A} = \begin{bmatrix} 0 & 0.25 & 0.25 & 0.25 & 0.25 \\ 0.5 & 0 & 0 & 0.5 & 0 \\ 0.33 & 0 & 0 & 0.33 & 0.33 \\ 1 & 0 & 0 & 0 & 0 \\ 0 & 0 & 0 & 1 & 0 \end{bmatrix}$$

Figure 11.2 – The transition probability matrix of our small "internet"

We learned previously that matrix **A** is usually called the cost matrix of the network but that it tends to be called a different name in different areas. Here, matrix **A** is typically called the transition probability matrix or importance matrix. It shows how PageRank transmits across all the outgoing links from each web page.

Each row of the transition probability matrix represents the proportion of the PageRank of a specific web page that will be transmitted to other web pages in future iterations, so they must add up to 1 to split the transmission of PageRank into parts making up the whole PageRank. Each column represents the incoming PageRank proportions transmitted from other web pages to a specific web page, so there is no need for them to add up to 1.

As we systematize these calculations into matrix algebra, we also need to represent the PageRanks as a vector. We will start with the initialized PageRanks all equal to $1/N = 0.5$. We can represent this as follows:

$$\mathbf{v}_0 = \begin{bmatrix} 0.2 \\ 0.2 \\ 0.2 \\ 0.2 \\ 0.2 \end{bmatrix}$$

Figure 11.3 – The initial PageRank vector

We'll define another matrix, **U**, which will actually be used in the PageRank calculations, and call it the update matrix, using the following formula:

$$\mathbf{U} = d\mathbf{A}^T + \frac{1-d}{N} \begin{bmatrix} 1 & \cdots & 1 \\ \vdots & \ddots & \vdots \\ 1 & \cdots & 1 \end{bmatrix}$$

Figure 11.4 – The matrix used in PageRank calculations

Recall from what we learned in *Chapter 6, Computational Algorithms in Linear Algebra*, that the *T* superscript indicates the transpose of the matrix where the rows are swapped with the columns, and we learned how to add and multiply matrices. We can replicate the PageRank formulas for all the websites in the next iteration at once as follows:

$$\mathbf{v}_i = \mathbf{U}\mathbf{v}_{i-1}$$

Let's write some Python code to apply this formula. Recall that the way we learned how to do matrix multiplication is to use NumPy for our small "internet" of five web pages, $N = 5$, and the default damping factor, $d = 0.85$:

```
# import the NumPy library
import numpy

# transition probability matrix
A = numpy.array([[0, 0.25, 0.25, 0.25, 0.25],
                 [0.5, 0, 0, 0.5, 0],
                 [0.33, 0, 0, 0.33, 0.33],
                 [1, 0, 0, 0, 0],
                 [0, 0, 0, 1, 0]])
```

```
# initialize the PageRank vector
v = numpy.array([[0.2], [0.2], [0.2], [0.2], [0.2]])

# the damping factor
d = 0.85

# the size of the "Internet"
N = 5

# compute the update matrix
U = d * A.T + (1 - d) / N

# compute the new PageRank vector
v = numpy.dot(U, v)

# print the new PageRank vector
print(v)
```

In this code, we first added the transition probability matrix and initialized the PageRank vector, damping factor, and size of the "internet." Then, we computed the update matrix. Finally, we computed the new PageRank vector after one iteration and printed it.

The output of the code is as follows:

```
[[0.3411]
 [0.0725]
 [0.0725]
 [0.3836]
 [0.1286]]
```

Rounding these values to two decimal places gives exactly the same as what we calculated by hand in *Chapter 5*, *Elements of Discrete Probability*, and replicated in the present chapter previously.

As we mentioned, it is common with PageRank implementations to run the PageRank update over and over until they stop changing:

```
# initialize the PageRank vector
v = numpy.array([[0.2], [0.2], [0.2], [0.2], [0.2]])

# print the initial vector
print('PageRank vector', 0, 'is', v.T)
```

```
# compute the PageRank vector for 15 iterations
for i in range(15):
    # compute the next PageRank vector
    v = numpy.dot(U, v)

    # round the PageRank vector to 3 places
    v = numpy.round(v, 3)

    # print the PageRank vector
    print('PageRank vector', i + 1, 'is', v.T)
```

This code initializes the PageRank vector, prints it, carries out the PageRank update for 15 iterations, rounds them, and prints each one. We print the transpose of the PageRank vector rather than the original simply so that the output does not take up too much space, which allows us to observe patterns in the evolution of the PageRanks more readily.

The output of the code is as follows:

```
PageRank vector 0 is [[0.2 0.2 0.2 0.2 0.2]]
PageRank vector 1 is [[0.341 0.073 0.073 0.384 0.129]]
PageRank vector 2 is [[0.408 0.102 0.102 0.264 0.123]]
PageRank vector 3 is [[0.326 0.117 0.117 0.293 0.145]]
PageRank vector 4 is [[0.362 0.099 0.099 0.305 0.132]]
PageRank vector 5 is [[0.359 0.107 0.107 0.289 0.135]]
PageRank vector 6 is [[0.351 0.106 0.106 0.296 0.136]]
PageRank vector 7 is [[0.356 0.104 0.104 0.295 0.134]]
PageRank vector 8 is [[0.354 0.105 0.105 0.293 0.135]]
PageRank vector 9 is [[0.353 0.105 0.105 0.294 0.134]]
PageRank vector 10 is [[0.354 0.105 0.105 0.293 0.134]]
PageRank vector 11 is [[0.353 0.105 0.105 0.293 0.134]]
PageRank vector 12 is [[0.353 0.105 0.105 0.293 0.134]]
PageRank vector 13 is [[0.353 0.105 0.105 0.293 0.134]]
PageRank vector 14 is [[0.353 0.105 0.105 0.293 0.134]]
PageRank vector 15 is [[0.353 0.105 0.105 0.293 0.134]]
```

As you can see, repeating the update over and over results in the PageRank vector converging to a certain set of numbers and not budging any further after about 10 iterations, as follows:

$$\mathbf{v}_{10} = \begin{bmatrix} 0.354 \\ 0.105 \\ 0.105 \\ 0.293 \\ 0.134 \end{bmatrix}$$

Figure 11.5 – The PageRank vector to which the PageRank updates converged

It should be stated that the calculations do continue to cause the PageRank vector to change, but the changes occur more than three places beyond the decimal point. So, practically speaking, we have found a steady state or equilibrium for the PageRanks.

What does all of this mean? If the five pages were returned by our search engine and we wanted to rank them and display the web pages sorted by PageRank from highest to lowest, our user would see web page W_1, followed by W_4, followed by W_5, followed by W_2 and W_3. We could break the tie at the bottom by realizing there is actually a difference in the PageRank if we look further beyond the decimal point. In a large-scale problem such as the actual internet, it would be incredibly unlikely to actually have equal PageRanks. However, if they were equal to the level of precision we have chosen to use for our computations, randomly breaking the tie would be practically fine.

As you have seen, we had to do a lot of computations to come to the preceding conclusion for a small five-web page "internet." Imagine the number of computations carried out by Google when they use the PageRank algorithm on the real internet and its billions of web pages before showing you the results for your query/search!

In this section, we learned about how the PageRank algorithm uses linear algebra to assign ranks to web pages returned by a web search. This allows the highly ranked web pages to be shown at the top of the list so that users can see the most relevant web pages first rather than having to go browsing through pages upon pages of irrelevant search results to find what they need.

We have applied the PageRank ideas to a small problem, but we will continue to build a realistic implementation of the PageRank algorithm in Python in the next section.

Implementing the PageRank algorithm in Python

In this section, we will take the insights we learned about the PageRank algorithm in the previous sections to write an effective Python implementation of the algorithm.

As we saw previously, the idea of the PageRank algorithm is to do some calculations to update the PageRank vectors over and over until they reach a steady-state PageRank vector. But we just ran it 15 times, looked at the numbers, and stopped when the updates become so small as to be insignificant.

However, there are a few obstacles to implementing this on a real, large-scale problem:

- If the "internet" of web pages is large, such as with the real internet, we could not really look at millions or billions of PageRanks in the updates and find when they have stopped changing.
- We cannot know in advance how many iterations we need to run for the PageRanks to converge to a steady state.
- We manually defined the initial state of the PageRank vector, which is impractical for a huge "internet."
- It depended specifically on the linking structure of the "internet" we considered, which would change in time in reality.
- We specified the size of the internet, N, and damping factor, d, manually.

In realistic implementations, we need to deal with all of these issues with our code:

- For problems 1–2, we need to find a way to automate the detection of when the PageRank algorithm converges to a steady state.
- For problem 3, we need to initialize the PageRank vector programmatically.
- For problems 4–5, we can write a function that takes the transition probability matrix $\mathbf{A}$ and damping factor d as inputs and finds the size of the "internet," N, from the matrix.

These latter two solutions are easy to implement, but solving the first two problems requires us to introduce some additional mathematics. The **Euclidean norm** of a vector is as follows:

$$\mathbf{v} = [v_1 \quad v_2 \quad \cdots \quad v_N]^T$$

It measures the length of a vector in an N-dimensional space and is computed as follows:

$$\|\mathbf{v}\| = \sqrt{\mathbf{v} \cdot \mathbf{v}} = \sqrt{v_1^2 + v_2^2 + \cdots + v_N^t}$$

Notice each component of the vector adds a positive value to the norm. If the components are near 0, then the norm will be near 0. If some of the components are large, then the norm will be large. This idea can be used to compute the distance between two vectors as well. If we define vector w similarly, the distance between vectors v and w is as follows:

$$\|\mathbf{v} - \mathbf{w}\| = \sqrt{(v_1 - w_1)^2 + (v_2 - w_2)^2 + \cdots + (v_N - w_N)^2}$$

So, here, if the $v_j - w_j$ differences are small, then the distance will small.

This mathematical tool gives us a way to measure how different vectors are. As we have learned, the PageRank algorithm is an iterative process that updates PageRank vectors over and over until they settle into a steady state with the following formula:

$$\mathbf{v}_i = \mathbf{U}\mathbf{v}_{i-1}$$

To solve problems 1–2 outlined previously, we can compute updates until the distance between vectors v_i and v_{i-1} is very small. Our function can actually accept an input of a small number, ε, called the error threshold and have the algorithm stop updating the PageRank vectors when the differences are smaller than this small error term, that is, when the following is true:

$$\|\mathbf{v}_i - \mathbf{v}_{i-1}\| < \varepsilon$$

This means we will need to save the PageRank vector from before each update so that we can find the difference between them.

Let's write some code implementing solutions to all of these problems by writing the PageRank algorithm as a function. It will be a little long, so we will break it down into parts and explain each part as we go. First, we write some documentation about the function:

```
# The PageRank algorithm for ranking search results
#
# INPUTS
# A - the transition probability matrix
# d - the damping factor, default = 0.85
# eps - the error threshold, default = 0.0005
# maxIterations - the maximum iterations it can run before
  # stopping
```

```
    # verbose - if true, the algorithm prints the progress of
    # PageRank
    #
    # OUTPUTS
    # vNew - the steady state PageRank vector
```

Next, we define the function, find the size of the "internet," and initialize several variables that we need:

```
def PageRank(A, d = 0.85, eps = 0.0005, maxIterations,
             verbose = False):
    # find the size of the "Internet"
    N = A.shape[0]

    # initialize the old and new PageRank vectors
    vOld = numpy.ones([N])
    vNew = numpy.ones([N])/N

    # initialize a counter
    i = 0
```

Then, we find the update matrix, **U**:

```
    # compute the update matrix
    U = d * A.T + (1 - d) / N
```

Then, we run the update over and over until the change in the PageRank vectors from iteration to iteration is sufficiently small:

```
    while numpy.linalg.norm(vOld - vNew) >= eps:
        # if the verbose flag is true, print the progress at
        # each iteration
        if verbose:
            print('At iteration', i, 'the error is',
                numpy.round(numpy.linalg.norm(vOld - vNew),
                    3), 'with PageRank', numpy.round(vNew, 3))

        # save the current PageRank as the old PageRank
        vOld = vNew

        # update the PageRank vector
```

```
            vNew = numpy.dot(U, vOld)

            # increment the counter
            i += 1
```

If the code does not converge within maxIterations, we will stop, print the error, and return the current PageRank vector and iteration:

```
# if it runs too long before converging, stop and notify the
# user
        if i == maxIterations:
            print('The PageRank algorithm ran for',
                   maxIterations, 'with error',
                   numpy.round(numpy.linalg.norm(vOld - vNew),
                   3))

            # return the PageRank vector and the
            return vNew, i
```

Finally, we return the steady-state PageRank vector and the number of iterations it took to converge:

```
    # return the steady state PageRank vector and iteration
      # number
    return vNew, i
```

Let's see whether it works with the example we used previously:

```
# transition probability matrix
A = numpy.array([[0, 1/4, 1/4, 1/4, 1/4],
                 [1/2, 0, 0, 1/2, 0],
                 [1/3, 0, 0, 1/3, 1/3],
                 [1, 0, 0, 0, 0],
                 [0, 0, 0, 1, 0]])

# Run the PageRank algorithm with default settings
PageRank(A)
```

The output is as follows:

```
(array([0.3565286 , 0.10584025, 0.10584025, 0.29600666,
   0.13578424]), 11)
```

We confirm this gives the same results we found previously in 11 iterations. So, what else can we do with the algorithm?

Suppose the structure of our small "internet" changes by changing the linking structure, which happens all the time in reality when people modify their web pages and create new web pages. Search engines periodically crawl the internet to find these changes.

If web page W_3 suddenly went viral and all the other web pages added links to it, this new "internet" could be represented by the following directed network:

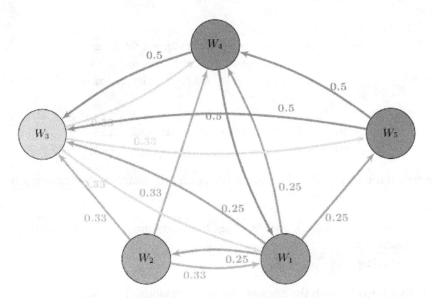

Figure 11.6 – A directed network representing an "internet" of five web pages after web page W_3 has gone viral and been linked on every other web page in the "internet"

Let's try PageRank again to see how the PageRanks will change due to web page W_3 becoming more popular. We will use the same calculations as previously but use a different transition probability matrix corresponding to the new state of our small internet:

```
# transition probability matrix
B = numpy.array([[0, 1/4, 1/4, 1/4, 1/4],
                [1/3, 0, 1/3, 1/3, 0],
                [1/3, 0, 0, 1/3, 1/3],
                [1/2, 0, 1/2, 0, 0],
                [0, 0, 1/2, 1/2, 0]])

# Run the PageRank algorithm with default settings
PageRank(B, verbose = True)
```

The output is as follows:

```
(array([0.2365497 , 0.08030807, 0.27603383, 0.24860661,
 0.15850179]), 8)
```

Since web page W_3 has increased in popularity, its PageRank should go up because it is more likely that users are looking for that web page. As you can see, its PageRank moved up from `0.105` to `0.276`.

In this section, we have written a realistic implementation of the PageRank algorithm, which takes in a transition probability matrix, initializes the PageRanks of each web page to the same proportion, and returns the steady-state PageRank as well as the number of iterations it took to converge.

In the next section, we will use our implementation on a much larger scale problem to see how well it works.

Applying the Algorithm to Real Data

Let's use our Python implementation of the PageRank algorithm to some larger-scale data. We will use a dataset shared by J. Kleinberg at Cornell by crawling the web to find web pages containing the word `California`. It is a text file in the following form:

```
Type Source Destination
n 0 http://www.berkeley.edu/
n 1 http://www.caltech.edu/
...
n 9663 http://www.cs.ucl.ac.uk/external/P.Dourish/hotlist.html
e 0 449
e 0 450
...
e 9663 7907
```

The first part contains 9,663 web pages that have the word `California`, and the rest is an adjacency list for the graph representing the "internet" of these 9,663 web pages. For example, take the following line:

```
e 0 499
```

This means web page `0` has a link to web page `499`. In order to implement PageRank on this dataset, we need to create an adjacency matrix.

Let's use some Python code to read this data file into a `pandas` DataFrame and display it:

```
# import the pandas library
import pandas

# read the txt file into a dataframe
data = pandas.read_csv("California.txt", delimiter=' ')

# display the dataframe
data
```

The output is as follows:

	Type	Source	Destination
0	n	0	http://www.berkeley.edu/
1	n	1	http://www.caltech.edu/
2	n	2	http://www.realestatenet.com/
3	n	3	http://www.ucsb.edu/
4	n	4	http://www.washingtonpost.com/wp-srv/national/...
...	...	...	...
25809	e	9663	1806
25810	e	9663	266
25811	e	9663	7905
25812	e	9663	70
25813	e	9663	7907

25814 rows × 3 columns

Next, we preprocess the data to extract the adjacency list, drop all the e strings in the first column, convert the remaining numerical portion into a NumPy array, and store the numbers as integers:

```
# preprocess the data

# select only the rows with type 'e'
adjacencies = data.loc[data['Type'] == 'e']

# drop the 'Type' column
adjacencies = adjacencies.drop(columns = 'Type')

# convert the adjacency list to a NumPy array
adjacencies = adjacencies.to_numpy()
```

```
# convert the adjacency list to integers
adjacencies = adjacencies.astype('int')

# print the adjacency list
print(adjacencies)
```

The output is as follows:

```
[[   0  449]
 [   0  450]
 [   0  451]
 ...
 [9663 7905]
 [9663   70]
 [9663 7907]]
```

Next, let's convert the adjacency list into an adjacency matrix:

```
# convert the adjacency list to an adjacency matrix

# find the number of webpages and initialize A
N = numpy.max(adjacencies) + 1
A = numpy.zeros([N, N])

# iterate over the rows of the adjacency list
for k in range(adjacencies.shape[0]):
    # find the adjacent vertex numbers
    i, j = adjacencies[k,]

    # put 1 in the adjacency matrix
    A[i, j] = 1
```

Next, we need to convert C into the transition probability matrix by dividing each 1 corresponding to an outgoing link by the total number of outgoing links from that web page. In other words, we divide each row by its row sum:

```
# convert A to the transition probability matrix

# divide each row of A by its row sum
rowSums = A.sum(axis = 1)[:,None]

# divide A by the rowSums
C = numpy.divide(A, rowSums, where = rowSums != 0)
```

Next, let's run PageRank on this transition probability matrix:

```
# run PageRank
v, i = PageRank(A)

# print the steady state PageRank vector and iteration number
print(v)
print(i)
```

The output is as follows:

```
[2.79688870e-05 6.29671046e-06 2.06171425e-07 ... 9.48337601e-
   08 9.48337601e-08 9.48337601e-08]
14
```

As we can see, feeding this large transition probability matrix of dimension 9,663 by 9,663 converges to a steady-state PageRank vector in 14 iterations.

We will then sort the PageRanks from highest to lowest and save the indices of the sorted list:

```
# sort the PageRanks in ascending order
ranks = numpy.argsort(v)

# find the PageRanks in descending order
ranks = numpy.flip(ranks)
```

Then, let's return the top 10 web pages containing the word `California`:

```
# return the URLs of the top few webpages
rankedPages = pandas.DataFrame(columns = ['Type', 'Source',
   'Destination'])

# add the top 10-ranked webpages
for i in range(10):
    row = data.loc[(data['Type'] == 'n')
                 & (data['Source'] == ranks[i])]
    rankedPages = rankedPages.append(row)

# display the top 10
rankedPages.drop(columns = ['Type', 'Source'])
```

A screenshot of the output is as follows:

	Destination
4391	http://search.ucdavis.edu/
1488	http://www.ucdavis.edu/
997	http://www.gene.com/ae/bioforum/
2408	http://www.lib.uci.edu/
8051	http://vision.berkeley.edu/VSP/index.shtml
1489	http://www.uci.edu/
718	http://www.students.ucr.edu/
211	http://spectacle.berkeley.edu/
17	http://www.calacademy.org/
4795	http://www.scag.org

Figure 11.7 – The top 10 web pages containing the word "California," ranked from highest to lowest PageRank

These results make quite a lot of sense. Rather than just returning web pages that say California frequently, these websites are all prominent entities in California. Most of the websites are from universities in the University of California system, which are all quite large universities that are linked by many other web pages. The last site is the Southern California Association of Governors, which is a metropolitan planning organization that provides large amounts of public data, meaning it is likely linked to by many web pages.

We have now applied our Python implementation of the PageRank algorithm to a real-world search example. Surprisingly, it converged very quickly, within a few seconds on a standard PC.

Summary

In this chapter, we learned about the PageRank algorithm developed in the late 1990s by the future founders of Google and their colleagues at Stanford. It revolutionized the world of search engines by providing an effective way to sort search results in such a way that much more relevant web pages to users' searches could be displayed at the top of the list.

We began by reviewing how search engines worked before PageRank, some prior innovations, and the general shortcomings of web search before PageRank.

Then, we moved on to applying a single PageRank update for a small "internet" of just five web pages introduced in *Chapter 5, Elements of Discrete Probability*. Instead of computing the formulas one by one by hand, we wrote a matrix form of the calculation and showed that it replicated the results from the previous chapter. We also learned that PageRank usually runs over and over until the PageRank vector converges to a steady state, which we did by running the updates for an arbitrary number of times until we saw it converge by inspection.

In the next section, we wrote a much better Python implementation of the PageRank algorithm, which detected convergence automatically by using a `while` loop that ran until the PageRank vectors on two successive iterations were sufficiently similar using the Euclidean norm. Next, we considered a scenario where one of the web pages in our small "internet" went viral and accumulated links from the other web pages, which resulted in the PageRank of this web page increasing because it had become a more likely landing spot for users.

Lastly, we brought in a large, real dataset of 9,663 web pages containing the word `California` and an adjacency list corresponding to links from one web page to another. We preprocessed the data to turn the adjacency list into an adjacency matrix. We further processed that into a transition probability matrix and ran the PageRank algorithm on this large example. It converged quickly and yielded some pretty intuitive results, ranking some prominent websites at the top.

In the next and final chapter of the book, we will learn about the method of **principal components analysis** (**PCA**), which is a method for reducing the dimensionality of data. This is quite an important task in machine learning.

12
Principal Component Analysis with Scikit-Learn

In this chapter, we will learn about **principal component analysis** (**PCA**), which is a core machine learning technique that reduces the dimensionality of large datasets to determine which variables can best explain strong patterns in data. We will first introduce some mathematical concepts about orthogonal matrices and bases. Then, we will explain the method and look at the scikit-learn library's implementation of PCA. Lastly, we will apply PCA to some real-world data.

In this chapter, we will cover the following topics:

- Understanding eigenvalues, eigenvectors, and orthogonal bases
- The principal component analysis approach to dimensionality reduction
- The scikit-learn implementation of PCA
- An application of PCA to real-world data

By the end of this chapter, you will have learned the intuition and mathematics behind PCA. You will also learn about the scikit-learn library's implementation of PCA and apply it to a real-world dataset.

Understanding eigenvalues, eigenvectors, and orthogonal bases

In this section, we will learn about the mathematical concepts behind PCA, such as eigenvalues, eigenvectors, and orthogonal bases. We will also learn how to find the eigenvalues and eigenvectors for a given matrix.

Many real-world machine learning problems involve working with a lot of feature variables; sometimes in the millions. This not only makes it harder for us to store the data due to its massive size but also leads to the slower training of machine learning models, making it harder for us to find an optimal solution. In addition, there is a chance that you are overfitting your model to the data. This problem is often referred to as the curse of dimensionality in the field of machine learning.

A solution to this curse of dimensionality is to reduce the dimensionality of datasets that have many feature variables. Let's try to understand this concept with the help of an example dataset: `pizza.csv`. This dataset can have 7 feature variables and 300 observations, which are categorized into 10 classes – pizza produced by 10 different pizza companies (companies A, B, C, D, E, F, G, H, I, and J). The original dataset along with the description can be found at `https://www.kaggle.com/shishir349/can-pizza-be-healthy`.

The columns of the dataset are as follows:

- **brand**: The names of the different pizza brands
- **moisture**: The water content per 100 grams of pizza
- **protein**: The protein content per 100 grams of pizza
- **fat**: The fat content per 100 grams of pizza
- **ash**: The ash content per 100 grams of pizza
- **sodium**: The sodium content per 100 grams of pizza
- **carbohydrates**: The carbohydrate content per 100 grams of pizza
- **calories**: The calorie content per 100 grams of pizza

Understanding eigenvalues, eigenvectors, and orthogonal bases   281

We will first import the dataset before we begin exploring it and later apply PCA to it to see how different pizza companies produce pizzas with different nutrient contents:

```
import pandas as pd
dataset = pd.read_csv('pizza.csv')
dataset.head()
```

We get the following table as the output when the preceding code is executed:

Brand	Moisture	Protein	Fat	Ash	....
A	27.82	21.43	44.87	5.11	....
A	28.49	21.26	43.89	5.34	....
A	28.35	19.99	45.78	5.08	....
A	30.55	20.15	43.13	4.79	....
....	....	....	....	....	....

(7 feature variables; 300 observations)

Figure 12.1 – Feature variables for the pizza dataset

We have seven feature variables in the preceding example; even though this sounds like a small number, it can have many correlated variables (*Figure 12.2*), which will increase the number of variables without adding much value. Dropping one or more of these features or combining our input variables will help in reducing the dimensions and make the problem more tractable. This is called feature elimination. A downside of this method is that we will eliminate any benefits that the dropped variables would have brought to our model. The following method is used to find the pairwise correlation between all the columns in a DataFrame:

```
dataset.corr()
```

The preceding code generates a correlation matrix as shown in the following figure:

	Moisture	Protein	Fat	Ash	....
Moisture	1.000	0.360	-0.171	0.266	....
Protein	0.360	1.000	0.498	0.824	....
Fat	-0.171	0.498	1.000	0.792	....
Ash	0.266	0.824	0.792	1.000	....
....	....	....	....	....	....

Figure 12.2 – Correlation matrix for the pizza dataset

A correlation matrix is a square and symmetrical matrix that has a value of 1 along its diagonal, which shows that each variable is perfectly correlated to itself. A correlation coefficient of 1 means that two variables are highly positively correlated, and a correlation coefficient value of -1 suggests that they are negatively correlated – when one value increases, the other decreases, and vice versa. Usually, it is hard to find variables are have a correlation coefficient of 1, so the closer it is to 1 or -1, the stronger the relationship, and vice versa.

The same variables are present in both the rows and columns of this matrix and show how each of the variables are correlated to another variable in the dataset. For example, the correlation coefficient between the protein and ash content is 0.824, which shows that they are highly positively correlated. This means that if a pizza sample has higher protein content, then it is more likely to have higher ash content, and vice versa.

Another way to achieve dimensionality reduction is to perform feature extraction on the dataset. In this method, we create new feature variables that are a combination of the original independent feature variables. We then rank these new feature variables based on how well they capture the variation in the original dataset. Here we have an option to only keep a certain number of the new feature variables, dropping the least important ones and still retaining the valuable parts of the old feature variables. PCA is used for this purpose and is a feature extraction algorithm.

Hence, the goal of dimensionality reduction is to reduce the number of feature variables while preserving as much information as possible for a dataset, which naturally comes at the expense of accuracy. This can be summed up as follows: increasing the simplicity of the machine learning model at the cost of a little reduction in accuracy.

Now that we have high-level knowledge of the problems that arise due to higher dimensions and know that the number of dimensions can be reduced, we will investigate the mathematical basis for these concepts.

An eigenvector v of a matrix **A** is a vector with a very special property: if you multiply the eigenvector by the matrix **A**, it maintains its original direction (the direction in which the vector was pointing initially before the matrix multiplication), as shown in *Figure 12.3* and *Figure 12.4*. However, the multiplication may squish, or stretch, and may reverse the eigenvector by a scalar factor. This scalar factor by which the eigenvector is squished or stretched is called an eigenvalue. The mathematical representation is shown here:

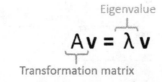

Figure 12.3 – Mathematical representation

By solving the preceding equation, we can find the value of the eigenvalues and the corresponding eigenvectors of **A**. The equation to solve to find the eigenvalues and eigenvectors is shown here:

$$Av = (\lambda I)v$$

$$(A - \lambda I)v = 0$$

Here *I* is the identity matrix and *v* is a non-zero vector.

Non-zero solutions exist only if $(A - \lambda I)$ is a singular matrix, which means that the determinant of $(A - \lambda I)$ is 0. Hence, the eigenvalues of *A* are roots of the polynomial $det(A - \lambda I)$. The number of eigenvalues is at most the number of dimensions. *Figure 12.3* shows a vector *v* in a two-dimensional X-Y coordinate plane:

$$\mathbf{v} = \begin{bmatrix} 1 \\ 1 \end{bmatrix}$$

The preceding vector is plotted as follows:

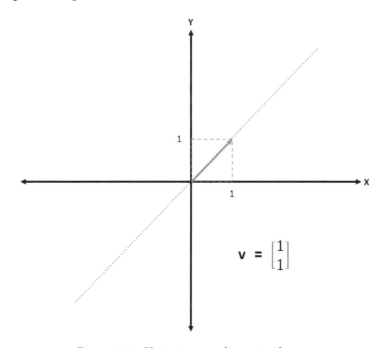

Figure 12.4 – Vector is a two-dimensional space

Now, let's apply a transformation **A** to the vector shown previously:

$$A = \begin{bmatrix} 1 & 1 \\ 1 & 1 \end{bmatrix} \begin{bmatrix} 1 \\ 1 \end{bmatrix}$$

$$Av = \begin{bmatrix} 1 & 1 \\ 1 & 1 \end{bmatrix} \begin{bmatrix} 1 \\ 1 \end{bmatrix} = \begin{bmatrix} 2 \\ 2 \end{bmatrix} = 2 \begin{bmatrix} 1 \\ 1 \end{bmatrix} = \lambda v$$

From the preceding computation, we can see that multiplying by A doubled the length of vector **v**, but the direction did not change, as shown in *Figure 12.5*:

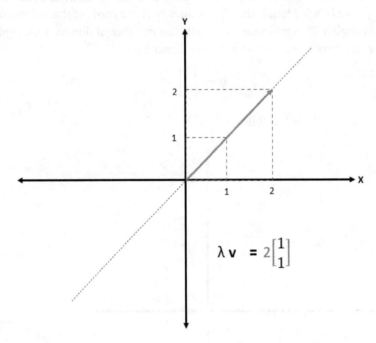

Figure 12.5 – Vector after undergoing transformation

Hence, we can say that the eigenvalue is 2 and the eigenvector is $\begin{bmatrix} 1 \\ 1 \end{bmatrix}$. This idea of eigenvalues and eigenvectors holds true for higher dimensions as well.

Let's try to find the eigenvalues and eigenvectors for a matrix **A** using Python:

$$A = \begin{bmatrix} 3 & 1 \\ 1 & 3 \end{bmatrix}$$

The code for finding eigenvalues and vectors for the preceding matrix is as follows:

```
import numpy as np
A = np.array([[3,1], [1,3]])
l, v = np.linalg.eig(A)
print("The eigenvalues are:\n ",l)
print("The eigenvectors are:\n ", v)
```

The output gives us the eigenvalues as well as the corresponding eigenvectors:

```
The eigenvalues are:
   [4. 2.]
The eigenvectors are:
   [[ 0.70710678 -0.70710678]
    [ 0.70710678  0.70710678]]
```

After going through the preceding exercise, you might ask a very important question: why do we need to transform vectors? Transformations are important since they can simplify problems by just rotating the axes so that we can have a simpler coordinate system to work with.

Let's now understand what orthogonal bases are. By the end of this section, you will have the basics that will be helpful while doing PCA for dimensionality reduction.

The standard basis of the d-dimensional space is made up of the following set of vectors:

$$\mathbf{e}_1 = \begin{bmatrix} 1 \\ 0 \\ \vdots \\ 0 \end{bmatrix}, \mathbf{e}_2 = \begin{bmatrix} 0 \\ 1 \\ \vdots \\ 0 \end{bmatrix}, \ldots, \mathbf{e}_d = \begin{bmatrix} 0 \\ 0 \\ \vdots \\ 1 \end{bmatrix}$$

Any datapoint in d-dimensional space $(c_1, c_2, \ldots, c_d)$ can be represented as

$$\begin{bmatrix} c_1 \\ c_2 \\ \vdots \\ c_d \end{bmatrix} = c_1 \mathbf{e}_1 + c_2 \mathbf{e}_2 + \cdots + c_d \mathbf{e}_d$$

if we choose to represent it in terms of the standard basis. This type of expression is called a linear combination of the vectors $e_1, e_2, \ldots, e_d$. In general, any set of d vectors that can construct *all* points in the d-dimensional space as linear combinations is called a basis.

The idea of PCA is to change the basis used to represent the points in a more efficient way, but we are not satisfied with just any basis. Notice that the standard basis has two special properties:

- **Property 1 (unit length)**: Every basis vector has a length of 1:

$$\|\mathbf{e}_i\| = \sqrt{\mathbf{e}_i \cdot \mathbf{e}_i} = 1$$

- **Property 2 (orthogonality)**: Every basis vector is orthogonal to every other basis vector, meaning that, if $i \neq j$, then the following is true:

$$\mathbf{e}_i \cdot \mathbf{e}_j = 0$$

Any basis with these two properties is called an orthonormal basis of the space. The idea of PCA is to change the basis we use to represent the covariance matrix from the standard basis to a special orthonormal basis made up of its *eigenvectors*. This basis is special because we can use the *eigenvalues* to rank the importance of the eigenvectors in representing the data, which allows us to reduce the size of the basis by deleting the vectors from it that have the least impact on the data.

In this section, we learned about mathematical basics such as eigenvalues, eigenvectors, and orthogonal bases, which we will use in the next section to understand dimensionality reduction with PCA.

The principal component analysis approach to dimensionality reduction

In this section, we will learn about the general idea of PCA and go through the steps for performing PCA on our dataset.

PCA is a method for reducing the dimensions of data by using some ideas from linear algebra to map the rows from a feature variable matrix **X** from its default d-dimensional space to an r-dimensional space for some $r < d$ by making use of principal components and the subsequent use of these components in understanding the data better.

From the previous section, we know that there are two types of dimensionality reduction methods: feature elimination and feature extraction. PCA falls into the latter category. It combines our input feature variables in a way that allows us to drop the least important variables (out of the new feature variables generated after performing PCA) while still retaining the valuable parts of all the input variables. In addition, the new feature variables after performing PCA are independent of one another.

The principal component analysis approach to dimensionality reduction 287

How do you decide on whether to apply PCA to your setup of input feature variables? This is an important question to answer before getting started with PCA:

- It is a good idea to consider PCA if you are working with a lot of variables and want to reduce them but are not sure about which variables are the least important ones.
- After applying PCA, the variables become less interpretable since PCA generates new feature variables that are a combination of the old variables.
- PCA makes sure that the variables generated are linearly independent of each other, making it easier to apply linear models such as linear regression to a dataset.

Let's try to understand PCA with some visuals before jumping into the details of it. The process of finding principal components starts with a covariance matrix, which is simply a correlation matrix that has not been normalized. Then we go on to find the eigenvalues and eigenvectors for this covariance matrix. This will give us an idea of the most important (principal) directions and how important these directions are – eigenvectors with higher eigenvalues are considered more important and they are ranked from most important to least important.

For our example dataset, let's say our scatterplot is shown in *Figure 12.6*. There are two main directions when it comes to the alignment of the datapoints – green and orange. As we can see, the variance in the data is more in the green direction as compared to the orange, making it more important:

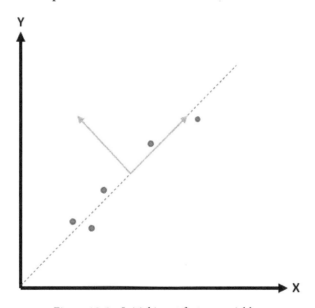

Figure 12.6 – Initial input feature variables

It would be easier to work with this dataset if we could transform our data to align with these important directions as shown in *Figure 12.7*. The data is transformed in such a way that the green and orange directions now align with our *x* and *y* axes. An important thing to keep in mind is that the principal directions are orthogonal to each other:

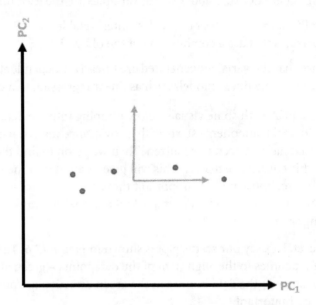

Figure 12.7 – Initial input feature variables after being transformed by PCA

It is important to keep in mind that even though the example shown here has two dimensions, this same principle applies to higher dimensions as well. PCA becomes even more important when we have a higher dimensional dataset since it helps us to project our dataset into lower dimensions while preserving most of the details captured by the original variables.

Let's now move on to looking at PCA with some more mathematical rigor. Assume that we have a dataset with *d*-dimensions (or columns) and *n* rows, where each of the columns corresponds to an independent variable:

- **Step 1**: It is important to standardize the dataset before moving ahead with PCA when different feature variables have different ranges. This makes sure that the mean for the variable is 0 and the variance is 1. The mathematical equation for normalizing a variable is shown here:

$$d_1(\text{normalized}) = \frac{(d_1 - \bar{d}_1)}{S}$$

Here d_1 is the original data value, $\bar{d}_1$ is the mean of the concerned feature variable, and S is the standard deviation.

We must apply this normalization for the entire dataset (all feature variables). Let's name this matrix C.

- **Step 2**: In this step, we take the transpose of matrix C and calculate the value for C^TC. This resulting matrix is called the **covariance matrix**. The covariance matrix shows how all the variables in a dataset are related to each other. The covariance matrix is similar to a correlation matrix, but not scaled to have entries between *-1* and *1*.

- **Step 3**: We then calculate the eigenvalues and vectors of this covariance matrix as shown in the previous section. Let's call the matrix containing the eigenvalues along the diagonal and zero everywhere else Λ and the matrix containing the eigenvectors V. These eigenvectors in V define the directions of the new axis, as shown in *Figure 12.7*. The corresponding eigenvalues determine the importance of each eigenvector and the information about the distribution of data that it carries:

$$\Lambda = \begin{bmatrix} \lambda_1 & \cdots & 0 \\ \vdots & \ddots & \vdots \\ 0 & \cdots & \lambda_d \end{bmatrix} \qquad \mathbf{V} = \begin{bmatrix} \mathbf{v}_1 & \mathbf{v}_2 & \cdots & \mathbf{v}_d \end{bmatrix}$$

$\lambda_1, \lambda_2, ..., \lambda_d$ are the eigenvalues. $v_1, v_2, ..., vd$ are the eigenvectors in the V matrix.

- **Step 4**: The eigenvalues are sorted in descending order and so are the eigenvectors. The sorted matrices are λ^* and V^*, respectively.

- **Step 5**: Next, we do matrix calculation to find $C^* = Cv$, where the observations of this new matrix are a combination of the original variables and the columns of C^* are linearly independent of one another. These new variables are not in an easily interpretable form, though.

- **Step 6**: This is the most crucial step, where we need to determine the number of components of C^* that we want to keep. This decision is based on the goal we are trying to achieve; if we want to plot a high-dimensional dataset in two dimensions, then we should keep the two most important principal components, and so on.

A more common way to make the decision is to calculate the proportion of variance explained by the selected principal components. Let's say you want your principal components to be able to explain 90% of the variation in the dataset. Then, you will have to include as many principal components as is required for the cumulative proportion of variance to reach your desired threshold of 90%.

The proportion of variance is obtained by dividing the sum of the eigenvalues of the selected features by the total sum of all eigenvalues of all features. Let's say that we selected the first two components out of *d* principal components; the proportion of variance would be as follows:

$$\frac{\lambda_1 + \lambda_2}{\lambda_1 + \lambda_2 + \cdots + \lambda_d}$$

To sum up, PCA helps us to reduce high-dimensional data to lower dimensions, and this can help with visualization, speeding up model training when it comes to machine learning, and more. We went through the various steps that are required to apply PCA to a dataset by making use of the eigenvalue, eigenvector, and orthogonal basis concepts that were covered in the first section. Lastly, we discussed the idea of the proportion of variance.

In the next section, we will learn about how to apply PCA to a dataset using the scikit-learn library.

The scikit-learn implementation of PCA

In this section, we will apply PCA to the `pizza.csv` dataset (which we explored in the first section of this chapter) using the scikit-learn library's `PCA` class.

As discussed in the previous section, there are two ways of choosing how many principal components to use, and the choice depends on the goal that you are trying to achieve – whether to reduce the dimensionality to plot something in 2-dimensional/3-dimensional space or keep enough principal components to achieve a certain proportion of variance.

First, we will implement the method where we can select the number of principal components we want to keep. We will reduce the 7-dimensional pizza dataset to two principal components so that we can visualize how the different pizzas produced by 10 different companies are different from each other when it comes to their nutritional content in a 2D plot instead of worrying about comparing and visualizing data in higher dimensions.

We will start by importing the dataset and then dropping the brand column from it. This is done to make sure that all our feature variables are numbers and hence can be scaled/normalized. We will then create another variable called `target`, which will contain the names of the brands of pizzas:

```
import pandas as pd
dataset = pd.read_csv('pizza.csv')
#Dropping the brand name column before standardizing the data
```

```
df_num = dataset.drop(["brand"], axis=1)

# Setting the brand name column as the target variable
target = dataset['brand']
```

Now that we have the dataset in order, we will normalize the columns of the dataset to make sure that the mean for a variable is 0 and the variance is 1. We will use StandardScaler, available in the scikit-learn library:

```
#Scaling the data (Step 1)
from sklearn.preprocessing import StandardScaler
scaler = StandardScaler()
scaler.fit(df_num)
scaled_data = scaler.transform(df_num)
```

After the data is scaled, we are ready to apply scikit-learn's PCA class to our dataset to obtain our principal components. We will restrict the number of principal components to two, which will enable us to later plot our principal components in a 2-dimensional plot:

```
#Applying PCA to the scaled data
from sklearn.decomposition import PCA

#Reducing the dimensions to 2 components so that we can have a
    # 2D visualization
pca = PCA(n_components = 2)
pca.fit(scaled_data)
#Applying to our scaled dataset
scaled_data_pca = pca.transform(scaled_data)
#Check the shape of the original dataset and the new dataset
print("The dimensions of the original dataset is: ", scaled_
    data.shape)
print("The dimensions of the dataset after performing PCA is:
    ", scaled_data_pca.shape)
```

Here is the output:

```
The dimensions of the original dataset are:   (300, 7)
The dimensions of the dataset after performing PCA is:   (300,
    2)
```

292 | Principal Component Analysis with Scikit-Learn

Now we have reduced our 7-dimensional dataset to its two principal components, which can be seen in the preceding dimensions. We will move forward with plotting the principal components to check whether two principal components were enough to capture the variability in the dataset – the different nutritional content of pizzas produced by different companies. We will be using the matplotlib library for the plotting:

```
#Plotting the principal components
import matplotlib.pyplot as plt
import seaborn as sns

sns.scatterplot(scaled_data_pca[:,0], scaled_data_pca[:,1],
   target)
plt.legend(loc="best")
plt.gca().set_aspect("equal")
plt.xlabel("Principal Component 1")
plt.ylabel("Principal Component 2")
plt.show()
```

Here is the output:

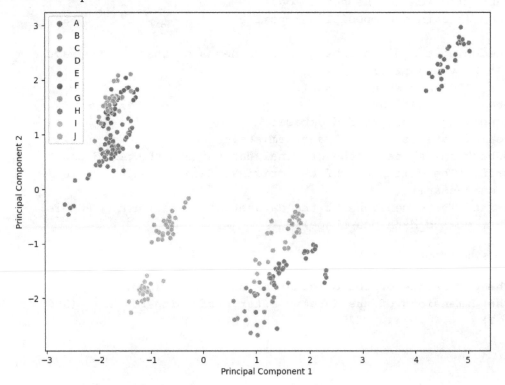

Figure 12.8 – Principal components of the pizza dataset

We plotted the first principal component against the second principal component and used the `target` column, which carried the names of the different pizza brands, to color the datapoints. We can clearly see the distinction between the pizzas produced by different pizza companies, with minor overlaps.

As mentioned previously, one of the demerits of PCA is that the plot is hard to interpret since the principal components are a combination of the original dataset.

Now, we will move on to perform PCA in a way where we do not choose the number of desired principal components; rather, we choose the number of principal components that add up to a certain desired variance. The Python implementation of this is very similar to the previous way with very slight changes to the code, as shown here:

```
import pandas as pd

dataset = pd.read_csv('pizza.csv')

#Dropping the brand name column before standardizing the data
df_num = dataset.drop(["brand"], axis=1)
# Setting the brand name column as the target variable
target = dataset['brand']

#Scaling the data (Step 1)
from sklearn.preprocessing import StandardScaler
scaler = StandardScaler()
scaler.fit(df_num)
scaled_data = scaler.transform(df_num)

#Applying PCA to the scaled data
from sklearn.decomposition import PCA

#Setting the variance to 0.95
pca = PCA(n_components = 0.95)
pca.fit(scaled_data)
#Applying to our scaled dataset
scaled_data_pca = pca.transform(scaled_data)
#Check the shape of the original dataset and the new dataset
print("The dimensions of the original dataset are: ", scaled_data.shape)
print("The dimensions of the dataset after performing PCA is: ", scaled_data_pca.shape)
```

Here is the output:

```
The dimensions of the original dataset are:    (300, 7)
The dimensions of the dataset after performing PCA is:    (300,
   3)
```

As we can see from the output, three principal components are required to capture 95% of the variance in the dataset. This means that by choosing two principal directions previously, we were capturing < 95% of the variance in the dataset. Despite capturing < 95% of the variance, we were able to visualize the fact that the pizzas produced by different companies have different nutritional contents.

In this section, we looked at the implementation of PCA using scikit-learn, which performs all the steps mentioned in the previous section under the hood and provides us with a quick result. We imported the dataset, dropped the non-numeric columns, and scaled each column of the dataset to make sure that the mean was 0 and the variance was 1. We then applied the PCA algorithm and selected the first two principal directions to visualize the dataset. We also performed a similar process but this time set the PCA algorithm to capture 95% of the variance in the dataset, for which it needed to capture three principal components.

In the next section, we will apply PCA to the popular MNIST dataset and analyze the number of principal components required to capture a required amount of variance in the dataset.

An application to real-world data

In this section, we will apply PCA to the MNIST dataset. The MNIST dataset is one of the most famous datasets in machine learning and contains handwritten digits that are used to train image processing algorithms. We will be using version 1 of the dataset, where each picture of every digit has 784 features. We will transform these features into a 28 x 28 matrix for visualization purposes. Each element of this matrix is a number between 0 (white) and 255 (black).

The first step is to import the data as shown in the following code. It is going to take some time since it is a big dataset, so hang tight. The dataset contains images of 70,000 digits (0-9), and each image has 784 features:

```
#Importing the dataset
from sklearn.datasets import fetch_openml
mnist_data = fetch_openml('mnist_784', version = 1)

# Choosing the independent (X) and dependent variables (y)
X,y = mnist_data["data"], mnist_data["target"]
```

Now that we have the dataset imported, we will move on to visualize an image of a digit to get familiar with the dataset. For visualization, we will use the matplotlib library. We will visualize the 50,000th digit image. Feel free to check out other digit images of your choice – make sure to use an index between 0 and 69,999. We will set colormap to binary to output a grayscale image, which is implemented in the following code:

```
#Plotting one of the digits
import matplotlib.pyplot as plt
plt.figure(1)
#Plotting the 50000th digit
digit = X[50000]
#Reshaping the 784 features into a 28x28 matrix
digit_image = digit.reshape(28,28)

plt.imshow(digit_image, cmap='binary')
plt.show()
```

Here is the output:

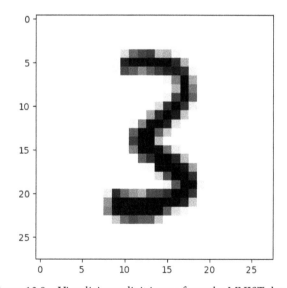

Figure 12.9 – Visualizing a digit image from the MNIST dataset

As you can see in the preceding figure, the image is *28 x 28* in size.

Next, we will move on to scaling the dataset by making using of `StandardScaler` to standardize the features by setting the mean to 0 and variance to 1:

```
#Scaling the data
from sklearn.preprocessing import StandardScaler
scaled_mnist_data = StandardScaler().fit_transform(X)
print(scaled_mnist_data.shape)
```

Here is the output:

```
(70000, 784)
```

`scaled_mnist_data` is a *70,000 x 784* matrix.

Now that we have our data in the form we want it to be in, we will go ahead and apply PCA to it. We will set the number of principal components to 784:

```
#Applying PCA to ur dataset
from sklearn.decomposition import PCA

pca = PCA(n_components=784)
mnist_data_pca = pca.fit_transform(scaled_mnist_data)
```

Now that we have the principal components figured, we will find the cumulative variance captured by these principal components. In other words, we will know how many principal components we need to consider to capture 90% of the variance in the original dataset. We will use the NumPy library to calculate the variance captured by each component and the cumulative variance. The equation for calculating the percentage variance captured by each component is as follows:

$$\text{Percentage variance explained by each PC} = \frac{\text{Variance explained by each PC}}{\text{Sum of variance explained by all PCs}}$$

Cumulative variance can be calculated by adding the variance explained by each component as we move from one component to another. We will calculate the cumulative variance using the numpy library, as shown here:

```
#Calculating cumulative variance captured by PCs
import numpy as np
variance_percentage = pca.explained_variance_/np.sum(pca.
    explained_variance_)

#Calculating cumulative variance
cumulative_variance = np.cumsum(variance_percentage)
```

We will now visualize the cumulative variance to see how many principal components are needed to explain 90% of the variance in the original dataset:

```
#Plotting cumalative variance
import matplotlib.pyplot as plt
plt.figure(2)
plt.plot(cumulative_variance)
plt.xlabel('Number of principal components')
plt.ylabel('Cumulative variance explained by PCs')
plt.grid()
plt.show()
```

Here is the output:

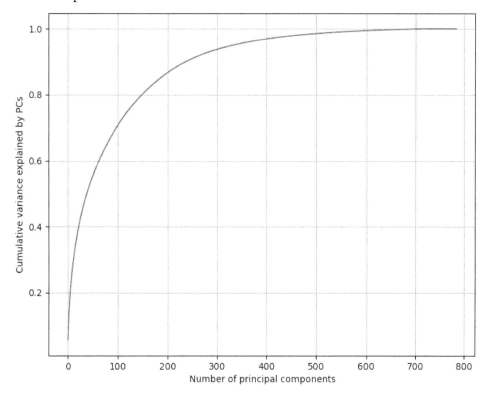

Figure 12.10 – Cumulative variance explained by principal components

From the preceding plot, we can see that a little less than 300 principal components (dimensions) are required to explain 90% of the variance in the original datasets.

After reducing the dimensions of the dataset, you can then move ahead with other machine learning algorithms (such as regression, clustering, and others) and apply them to your dataset. You should see a considerable decrease in the model training time with principal components.

In this section, we learned about applying PCA to the MNIST dataset and saw how we can use just 300 dimensions to capture a very high variance in the dataset. This is helpful when it comes to large datasets to reduce the disk space that is required for storage, reduce the computation time, and have a lower chance of model fitting.

Summary

In this chapter, we learned about eigenvalues, eigenvectors, and orthogonal bases and how these concepts connect to form a basis for dimensionality reduction. We then learned about the two types of dimensionality reduction methods – feature elimination and feature extraction. We discussed the different steps of performing Principal Component Analysis which falls into the feature extraction category for dimensionality reduction. We used the implementation of PCA from scikit-learn to apply the algorithm to our dataset, where we reduced the features in our pizza dataset from 7 to 2 and visualized the data. We were able to easily tell that the nutrients present in the pizzas manufactured by different companies were different. Lastly, we applied PCA to the MNIST dataset and figured out that only 300 principal components were needed to capture 90% of the variance in the dataset, as compared to the 784 feature variables that we had originally, reducing the dimensionality by more than 50%!

Congratulations! We have reached the end of our discrete mathematics with applications in Python journey. If you have followed along with all the mathematical concepts and the Python code snippets, you are in a very good position to carry forward this knowledge and apply it to more complex real-world problems. A suggestion would be to start working on new projects involving different datasets and showcasing your work on different platforms such as GitHub and Kaggle. This will not only help you showcase your achievements but also aid your understanding of the concepts better. Keep practicing! In the words of Confucius, "I hear and I forget. I see and I remember. I do and I understand."

Other Books You May Enjoy

If you enjoyed this book, you may be interested in these other books by Packt:

Applying Math with Python

Sam Morley

ISBN: 978-1-83898-975-0

- Get familiar with basic packages, tools, and libraries in Python for solving mathematical problems
- Explore various techniques that will help you to solve computational mathematical problems
- Understand the core concepts of applied mathematics and how you can apply them in computer science
- Discover how to choose the most suitable package, tool, or technique to solve a certain problem
- Implement basic mathematical plotting, change plot styles, and add labels to the plots using Matplotlib
- Get to grips with probability theory with the Bayesian inference and Markov Chain Monte Carlo (MCMC) methods

Hands-On Mathematics for Deep Learning

Jay Dawani

ISBN: 978-1-83864-729-2

- Understand the key mathematical concepts for building neural network models
- Discover core multivariable calculus concepts
- Improve the performance of deep learning models using optimization techniques
- Cover optimization algorithms, from basic stochastic gradient descent (SGD) to the advanced Adam optimizer
- Understand computational graphs and their importance in DL
- Explore the backpropagation algorithm to reduce output error
- Cover DL algorithms such as convolutional neural networks (CNNs), sequence models, and generative adversarial networks (GANs)

Leave a review - let other readers know what you think

Please share your thoughts on this book with others by leaving a review on the site that you bought it from. If you purchased the book from Amazon, please leave us an honest review on this book's Amazon page. This is vital so that other potential readers can see and use your unbiased opinion to make purchasing decisions, we can understand what our customers think about our products, and our authors can see your feedback on the title that they have worked with Packt to create. It will only take a few minutes of your time, but is valuable to other potential customers, our authors, and Packt. Thank you!

Index

Symbols

3-by-3 linear system
  example 131-134
10-by-10 linear system
  example 135, 136

A

acyclic graphs 173
adjacency data
  efficient storage 186
adjacency list
  about 181
  example 182
adjacency matrices 193
adjacency matrix
  about 182
  example 182, 183
  features 183
  for directed graph 184
  storing, in Python 185, 186
adjacent vertices 177
algorithms
  about 140
  computational complexity 140-144
  feasibility 142
  finiteness 141
  input 141
  output 141
  unambiguous 142
algorithms, from data structure
  delete 140
  insert 140
  search 140
  sort 140
  update 140
algorithms, performance
  fixed part 143
  space requirement 143
  time requirement 143
  variable part 143
analysis of algorithms 6
AND operator 52, 53
argument 21

B

base-6 number
  decimal value 48
base-n numbers
  about 46
  converting, to decimal numbers 48

decimal number 46
decimal numbers, example 46
defining 47
bases
  converting between 47
Bayesian spam filtering 97, 98
Bayes' theorem 93-97
best-fit lines 238
biconditional 25
Big-O Notation 145
binary numbers
  about 51, 52
  applications 51, 52
binary search algorithm
  about 161-164
  average case 164
  best case 163
  working 164
  worst case 163
binomial coefficient 72
binomial RV 99
Boolean algebra
  about 52
  AND operator 52, 53
  NOT operator 55, 56
  OR operator 54, 55
Boolean operators
  example 56-58
Brute Force
  used, for searching shortest
    path 212-214
brute-force algorithms
  efficacy 76
bytes 52, 67

C

Caesar cipher
  example 76-79
cardinality 12
Cartesian coordinate plane 111
Cartesian product
  about 66
  cardinality, of finite sets 66, 67
  for n sets 67
colors on computers
  example 68
combination
  about 71
  of set 72
combinations of balls
  example 73
combination, versus permutation
  for simple set 71
Comma-Separated Value (CSV) 57
common classes of computational
    complexity 164-166
commutative law 52
complexity of algorithms
  with fundamental control structures 151
complexity of complex functions 148-150
conclusion 21
conditional 25
conditional probability 93, 94
conjunction 24
connected components 177, 178
connected graphs 177, 178
consistent system
  about 118
  in RREF 129
constant complexity O(c) 145, 146
constants 150, 151

contradiction, using for
    mathematical proofs
  about 34, 35
  examples 35-38
contrapositive 30, 33
control structures
  about 151
  repetitive flow 155
  selection flow 153
  sequential flow 152
  using, in complexity of algorithms 151
converse 27
counting rule 66, 67
covariance matrix 289
cryptography 5
cycles 172, 173

D

damping factor 103
dataset
  about 236-238
  columns 236, 237, 280
decimal numbers
  base-n numbers, converting to 48
  converting, to base-2 (binary) 48, 49
  converting, to binary in Python 50
  converting, to hexadecimal in Python 50
  example 46
degree, of vertex
  about 171
  examples 171
  theorem 171
De Morgan's laws
  about 10, 11, 29, 30
  example 12

dependent system
  about 118
  in RREF 130
depth-first search (DFS)
  about 197, 199-201
  Python implementation of 201-204
digraph 175
Dijkstra's algorithm
  about 215
  applying, to small problem 216-221
  pattern 215
  used, for searching shortest
    path 214, 215
dimensionality reduction
  principal component analysis
    (PCA), approach 286-290
directed acyclic graphs (DAGs) 179
directed edges 175
directed graph
  about 175
  adjacency matrix 184
directed networks
  about 176
  example 176
discrete mathematics 4, 5
discrete mathematics, real-
    world applications
  about 5
  analysis of algorithms 6
  cryptography 5
  logistics 5
  machine learning 5
  relational databases 6
discrete probability
  basics 84
  elementary properties of probability 87
  events 85
  Laplacian probability 90

monotonicity theorem 88
outcomes 85
Principle of Inclusion-Exclusion 89
probability measure 86
random experiment 84
sample spaces 85
disjoint set 10
disjunction 25
domains 13
dot product
   of vectors 124

E

edges 169
eigenvalues bases 280-286
eigenvectors bases 280-286
Electronic Numerical Integrator
   and Computer (ENIAC) 51
elementary algebra 14
elementary properties of probability
   about 87
   example 87, 88
empty set 7
Euclidean norm 268
events 85
expectation
   about 100
   empirical random variable 101

F

factorials 69, 70
features 239
for loop 155-158
formal logic
   by truth tables 20
   cores ideas 24
   terminology 20, 21
functions
   about 13
   in elementary algebra 14
   versus relations 13

G

Gaussian elimination
   about 127-131
   linear systems, solving 128
Google PageRank I 102-105
Google's PageRank algorithm 260-267
graphs
   about 170
   degrees, of vertices 190, 191
   feature extraction 190
   searching 198
   storage 181
   using 178-181
greedy algorithm 215

H

hexadecimal numbers
   about 59, 60
   advantages 64
   applications 59, 60
   colors, defining on web 63, 64
   error messages, displaying 62
   locations, defining in computer
      memory 60, 61
   MAC addresses 62

I

if-elif-else conditionals 154
implication 25

inconsistent and dependent
    systems, NumPy
  example 134, 135
inconsistent system
  about 118
  in RREF 129
intractable problem 165
invalid 21

K

k-permutations
  of set 70, 71

L

Laplacian probability
  about 90
  calculating 90
  independent events 91
  tossing many coins, example 91, 93
  tossing multiple coins, example 91
Law of Total Probability 96
leading coefficient (pivot) 128
least-squares curves
  using, with NumPy 249-251
  using, with SciPy 249-251
least-squares lines
  using, with NumPy 245-249
least-squares method 238, 243-245
least-squares surfaces
  using, with NumPy 252-254
  using, with SciPy 252-254
Linear complexity O(n) 146, 147
linear equations
  example 112
  in two variables 110
linear relationship 238

linear search algorithm,
    about 160, 161
  average case 161
  best case 161
  working 161
  worst case 161
linear system of two equations,
    in two variables
  about 113
  consistent system 113
  dependent system 116, 117
  inconsistent system 114
linear systems
  matrix representations 119
  solving, with Gaussian elimination 128
  solving, with NumPy 133
linear systems of equations 110
line of best fit 240-243
logical connectives 24
logistics 5

M

machine learning 5
mathematical functions
  versus Python functions 15, 16
mathematical induction
  proofs 38
  proofs, example 39-44
mathematical proofs
  about 31
  examples 31-33
matrices
  about 119, 120
  addition 121
  in Python 193, 194
  multiplying 125, 126

scalar multiplication 122-124
subtraction 121
matrix multiplication 124, 125
mean 100
memory allocation
  applications 74
memory pre-allocation
  example 74, 75
minimum-edge paths, between v_i and v_j
  about 194
  example 194, 195
minimum spanning tree (MST) 180
MNIST dataset
  principal component analysis (PCA), applying, to 294-298
model parameters 239
monotonicity theorem 88
multiplication rules 95

N

negation 24
Network Interface Card (NIC) 62
networks
  about 170, 174
  shortest paths on 205
  storage 181
  using 178-181
nodes 170
non-connected graph
  example 183
non-deterministic polynomial time (NP) 165
Notation 145
NOT operator 55, 56
number of paths
  counting, between vertices of specified length 191, 192

NumPy
  least-squares curves, using with 249-251
  least-squares lines, using with 245-249
  least-squares surfaces, using with 252
  linear systems, solving 133

O

ordered pairs 175
OR operator 54, 55
orthogonal bases 280-286
outcomes 84, 85

P

packages installing, in Python
  reference link 57
PageRank 103
PageRank algorithm
  applying, to Real Data 273-277
  implementing, in Python 268-273
paths 172
permutation
  about 68
  of set 69
  of simple set 68
  playlists example 69
polynomial regression 249
polynomial time (P) 164
premises 21
principal component analysis (PCA)
  about 279
  applying, to MNIST dataset 294-298
  approach, to dimensionality reduction 286-290
  scikit-learn implementation 290-294
Principle of Inclusion-Exclusion 89
probability measure 86

proofs
  by truth tables 20
propositions 24
Python
  adjacency matrix, storing in 185, 186
  matrices powers 193, 194
  PageRank algorithm,
      implementing in 268-273
  used, for converting decimal
      numbers to binary 50
  used, for converting decimal
      numbers to hexadecimal 50
  weight matrix, storing in 188, 189
Python functions
  about 160
  versus mathematical functions 15, 16
Python implementation
  of depth-first search (DFS) 201-204
Python Implementation, of
    Dijkstra's algorithm
  about 221-225
  example 225-230
Python programming language
  logical conditions 153

Q

Quadratic complexity O(n2) 147, 148

R

random experiment
  about 84
  tossing coins example 85
  tossing multiple coins example 85, 86
random variables (RV)
  about 98, 99

data transfer errors example 99
empirical random variable 100
ranges 13
Real Data
  PageRank algorithm,
      applying to 273-277
red, green, and blue (RGB) 68
reduced row echelon form (RREF)
  about 128
  consistent system 129
  dependent system 130
  inconsistent system 129
regression 238-240
relational databases 6
relations
  about 13
  versus functions 13
repetitive flow 155
residuals 244

S

sample spaces 85
scikit-learn implementation
    of principal component analysis
      (PCA) 290-294
SciPy
  least-squares curves, using with 249-251
  least-squares surfaces, using
      with 252-254
SciPy library
  reference link 73
search algorithm, complexity
  about 159
  binary search algorithm 161-164
  linear search algorithm 160, 161
Search Engines
  developing, over time 258, 259

selection flow 153
sequential flow 152
set
  about 6
  basic operations 8, 10
  cardinality 12
  disjoint set 10
  elements 6
  empty set 7
  even and odd numbers, example 10
  examples 7
set-builder notation
  about 7
  using 8
set theory 6
Shortest-Distance Paths 206, 207
shortest path
  on networks 205
  problems 205
  problems, variations 205
  searching, with Brute Force 212-214
  searching, with Dijkstra's
    algorithm 214, 215
shortest path problem
  checking, whether solutions
    exist 208-211
  statement 207, 208
standard deviation 101
subset 7
sum of squared errors (SSE) 243-245
superset 7
symmetric matrix 183
systems of linear equations
  about 118
  solutions 118

T

teambuilding
  example 72, 73
temperatures and precipitation
    example 94, 95
terminology, for formal logic
  examples 22-26
timeit library
  reference link 144
tractable problem 164
transitivity law 28
traveling salesman problem
  example 79, 81
tree data structures
  searching 198
trees
  about 170, 173
  using 178-181
trendline 240
truth tables
  about 26
  formal logic by 20
  proofs by 20
truth tables, example
  contrapositive 30, 31
  De Morgan's laws 29, 30
  transitivity law, of conditional logic 28
truth tables, examples
  converse 27
types, repetitive flow
  for loop 155-158
  while loop 159

V

valid 21
variable 238
variance
  about 101
  empirical random variable example 102
  practical calculation 102
vectors 119
  dot product 124
vertices 169

W

weight matrix
  storing, in Python 188, 189
weight matrix, of directed network
  about 188
  examples 188
weight matrix, of network
  about 187
  examples 187
while loop 158, 159

Printed in the USA
CPSIA information can be obtained
at www.ICGtesting.com
LVHW070503201223
766892LV00005B/340